贡宝拉格常见植物资源彩色图谱

◎ 闫伟红　张晓庆　于凤春　马玉宝　著

中国农业科学技术出版社

图书在版编目（CIP）数据

贡宝拉格常见植物资源彩色图谱 / 闫伟红，张晓庆，于凤春，马玉宝 著 .—北京：中国农业科学技术出版社，2017. 12

ISBN 978-7-5116-3299-9

Ⅰ. ①贡… Ⅱ. ①闫… ②张… ③于… ④马… Ⅲ. ① 植物资源—锡林郭勒盟—图谱 Ⅳ. ① Q948.522.62-64

中国版本图书馆 CIP 数据核字（2017）第 251893 号

项目资助：

国家青年自然基金“不同纬度黄花苜蓿表观遗传多样性和低温下DNA表观遗传变化分析（31302017）”项目

合作单位：内蒙古锡林郭勒盟太仆寺旗草原工作站

Funded project:

National Natural Science Foundation of China “Epigenetic diversity of Medicago falcata L. at different latitudes and DNA analysis of epigenetic changes under low temperature stress (31302017)”

Cooperation Unit:

Taipus Temple Grassland Stations of Xilin Gol of Inner Mongolia

责任编辑 李冠桥
责任校对 马广洋

出 版 者 中国农业科学技术出版社
北京市中关村南大街12号　　邮编：100081
电　　话 （010）82109705（编辑室）（010）82109702（发行部）
（010）82109709（读者服务部）
传　　真 （010）82106625
网　　址 http: // www.castp.cn
经 销 者 全国各地新华书店
印 刷 者 北京富泰印刷有限责任公司
开　　本 880mm × 1 230mm　1/16
印　　张 10.25
字　　数 274千字
版　　次 2017年12月第1版　　2018年1月第2次印刷
定　　价 49.00元

《贡宝拉格常见植物资源彩色图谱》

著者名单

主　　著：闫伟红　张晓庆　于凤春　马玉宝

参　　著：（按姓氏笔画排序）

于林清　王照兰　王慧萍　宁　发　白　顺

田青松　刘　磊　祁　娟　江中东　吉木色

师文贵　吴洪新　张冬梅　陈立波　孟庆伟

李惠敏　杨　忠　赵杏花　高天明　催艳伟

摄　　影：白　顺

Editors: Yan Weihong　ZhangXiaoqing　Yu Fengchun　MaYubao

Assistant Editors:（Arranged by Family Name）

Yu Linqing　Wang Zhaolan　Wang Huiping　Ning Fa

Bai Shun　Tian Qingsong　Liu Lei　Qi Juan

Jiang Zhongdong　Ji Muse　Shi Wengui　Wu Hongxin

Zhang Dongmei　Chen Libo　Meng Qingwei

Li Huimin　Yang Zhong　Zhao Xinghua

Gao Tianming　Cui Yanwei

Photographer: Bai Shun

前　言

草地植物是宝贵的可再生资源，在维持天然草地生态平衡，发展草地畜牧业生产中，具有重要作用。

作者经过4年的野外调查，采集植物标本和拍摄图片，记录植物生境，对内蒙古锡林郭勒草原太仆寺旗贡宝拉格苏木范围内的植物种类、地理分布等，作了分类学研究，在此基础上编撰了《贡宝拉格常见植物资源彩色图谱》一书。

《贡宝拉格常见植物资源彩色图谱》一书收集了贡宝拉格草原常见植物111种，采用中、英文2种文字对31科111种植物的特征、生境、用途进行了描述，并刊载了大量彩色图谱，具有直观性和适用性，是了解贡宝拉格草原植物资源的工具书。本书可供牧草学、草原学、农、牧、林科技工作者及生产、教学、科研、医药学等部门参考。

苏木草原植物较多，本书收录并不全面。由于时间、条件和水平所限，难免有误，谨希广大读者、学者和同行批评指正。

在野外调研和本书编写期间，曾得到锡林格勒盟太仆寺旗草原站、内蒙古农业大学和内蒙古草原勘察设计院的相关领导和科研人员的支持和帮助，在此一并表示诚挚的感谢。

著　者

2017年11月

目　录

① 植株 ② 枝条上部 ③ 枝条下部

木贼科　Equisetaceae

木贼属　*Equisetum* L

问　荆　*Equisetum arvense* L.

〔特征〕

中小型植物。根茎斜升，直立和横走，黑棕色，节和根密生黄棕色长毛或平滑无毛。地上枝当年枯萎。枝二型。能育枝春季先萌发，高5～35cm，中部直径3～5mm，节间长2～6cm，黄棕色，无轮茎分枝，脊不明显，要密纵沟；鞘筒栗棕色或淡黄色，长约0.8cm，鞘齿9～12枚，栗棕色，长4～7mm，狭三角形，鞘背仅上部有一浅纵沟，孢子散后能育枝枯萎。不育枝后萌发，高达40cm，主枝中部直径1.5～3mm，节间长2～3cm，绿色，轮生分枝多，主枝中部以下有分枝。脊的背部弧形，无棱，有横纹，无小瘤；鞘筒狭长，绿色，鞘齿三角形，5～6枚，中间黑棕色，边缘膜质，淡棕色，宿存。侧枝柔软纤细，扁平状，有3~4条狭而高的脊，脊的背部有横纹；鞘齿3～5个，披针形，绿色，边缘膜质，宿存。孢子囊穗圆柱形，长1.8～4cm，直径0.9～1.0cm，顶端钝，成熟时柄伸长，柄长3～6cm。

〔生境〕

生于河道沟渠旁、疏林、荒野和路边、潮湿的草地、沙土地、耕地、山坡及草甸等处。

〔用途〕

药用，有止血、止咳、利尿、明目等功效。

〔Features〕

Small and medium-sized plant. Rhizomes obliquely ascending, erect and amphitropous, dark brown, internodes and roots covered with yellowish brown hairs or glabrous. Branches above ground withered the same year. Branches dimorphic. Fertile branches sprout in spring, 5 ~ 35cm tall, 3 ~ 5mm in diameter, internodes 2 ~ 6cm long, yellowish brown, no stem branches, ridges unconspicuous, with longitudinal furrows; sheath tube chestnut or light yellow, about 0.8cm long, sheath teeth 9 ~ 12, chestnut, 4 ~ 7mm long, narrowly triangular, a shallow longitudinal furrow on the top of sheath back, fertile branches withered after spore scattered. Sterile branches sprout later, up to 40cm tall, the middle part of main branches 1.5 ~ 3mm in diameter, internodes 2 ~ 3cm long, green, verticillate, branched below the middle part of main branches. The back of ridges is arc-shaped, non-edge, striate, not tuberculate; sheath tube narrow and long, green, sheath teeth triangular, 5 ~ 6, the middle part black brown, membranous at the margin, light brown, persistent. Lateral branches soft and slender, flat, with 3 ~ 4 narrow and tall ridges, the back of ridges striate; sheath teeth 3 ~ 5, lanceolate, green, membranous at the margin, persistent. Fertile spike cylindrical, 1.8 ~ 4cm long, 0.9 ~ 1.0cm in diameter, obtuse at the apex, petioles elongate when mature, 3 ~ 6cm long.

〔Habitat〕

River channels, open forests, the wilds and roadside, wet lawns, sand land, farmlands, hillside and meadows, etc.

〔Use〕

Medicinal use, hemostatic, cough relieving, diuretic and improving eyesight.

① 植株 ② 叶、茎 ③ 枝条

蓼 科 Polygonaceae L.

蓼 属 *Polygonum*

萹 蓄 *Polygonum aviculare* L.

〔特征〕

一年生草本，高10～40cm。茎平卧或上升，自基部分枝，有棱角。叶有极短柄或近无柄；叶片狭椭圆形或披针形，长1.5～3cm，宽5～10mm，顶端钝或急尖，基部楔形，全缘；托叶鞘质，下部褐色，上部白色透明，有不明显脉纹。花腋生，1～5朵簇生叶腋，遍布于全植株；花梗细而短，顶部有关节；花被5深裂，裂片椭圆形，绿色，边缘白色或淡红色；雄蕊8；花丝较短；子房卵形，花柱3。瘦果卵形，有3棱，黑色或褐色，生不明显小点，无光泽。

〔生境〕

生于田野、荒地、路边、水边湿地或沙地。

〔用途〕

药用，具有清热、利尿、解毒功效。

〔Features〕

Annual herb, 10 ~ 40cm tall. Stems prostrate or ascending, branched from the base, carinal. Leaves have very short stems or nearly impedicellate; leaf-blades narrowly oval or lanceolate, 1.5 ~ 3cm long, 5 ~ 10mm wide, obtuse or acuminate at the apex, cuneate at the base, the margin entire; ocrea, brown on the lower part, white transparent at the top, inconspicuous veinings. Flowers axillary,1 ~ 5-flowered clustered in axils, spreading all over the entire plant; peduncles thin and short, articulated at the top; perianths 5 parted, lobes oval, green, white or light red at the margin; stamens 8; filament short; ovary ovoid, stylus 3. Achenes ovate, trigonous, black or brown, with inconspicuous dots, lackluster.

〔Habitat〕

Fields, wastelands, roadsides, waterfronts, wetlands or sand lands.

〔Use〕

Medicinal use, clearing heat, diuretic and detoxicating.

① 植株 ② 植株 ③ 花 ④ 叶、花 ⑤ 花序 ⑥ 茎、叶

叉分蓼 *Polygonum divaricatum* L.

〔特征〕

多年生草本，茎直立，高70～120cm，无毛，自基部分枝，分枝呈叉状，开展，植株外形呈球形。叶披针形或长圆形，长5～12cm，宽0.5～2cm，顶端急尖，基部楔形或狭楔形，边缘通常具短缘毛，两面无毛或被疏柔毛；叶柄长约0.5cm；托叶鞘膜质，偏斜，长1～2cm，疏生柔毛或无毛，开裂，脱落。花序圆锥形，分枝开展；苞片卵形，边缘膜质，背部具脉，每苞片内具2～3花；花梗长2～2.5mm，与苞片近等长，顶部具关节；花被5深裂，白色，花被片椭圆形，长2.5～3mm，大小不相等；雄蕊7～8，比花被短；花柱3，极短，柱头头状。瘦果宽椭圆形，具3锐棱，黄褐色，有光泽，长5～6mm，超出宿存花被约1倍。花期7—8月，果期8—9月。

〔生境〕

生于山坡草地、山谷灌丛。

〔用途〕

药用，具有清热、消积、散瘿、止泻功效。

〔Features〕

Perennial herb, stems erect, 70 ~ 120cm tall, glabrous, branched from the base, branches forked, squarrose, appearance of plant spherical. Leaves lanceolate or long elliptic, 5 ~ 12cm long, 0.5 ~ 2cm wide, acuminate at the apex, cuneate or narrowly cuneate at the base, usually short tricholoma at the margin, glabrous or sparse pubescence on both sides; petioles about 0.5cm long; ocrea membranous, inclined, 1 ~ 2cm long, sparsely pubescence or glabrous, cracked, deciduous. Inflorescence coniform, branches squarrose; bracts ovate, membranous margins, dorsal surface veined, each bract has 2 ~ 3 flowers; peduncles 2 ~ 2.5mm long, approximately equal in length to bracts, articulated at the top; perianths 5-parted, white, tepals oval, 2.5 ~ 3mm long, sizes unequal; stamens 7 ~ 8, shorter than perianths; stylus 3, extremely short, stigma capitellate. Achene broadly oval, 3 sharp edges, yellowish-brown, lustrous, 5 ~ 6mm long, about 1 time longer than persistent perianths. Flowering July to August, and fruiting August to September.

〔Habitat〕

Hillsides, grasslands, valleys and bushwoods.

〔Use〕

Medicinal use, clearing heat, removing food retention, mitigating the goitre, and antidiarrheic.

① 植株　② 叶　③ 花序

西伯利亚蓼　*Polygonum sibiricum* Laxm.

〔特征〕

多年生草本，有细长的根状茎。茎斜升上或近直立，高6～20cm，通常自基部分枝。叶有短柄；叶片矩圆形或披针，近肉质，无毛，长5～8cm，宽5～15mm，顶端急尖，基部戟形或楔形，全缘。花序圆锥状，顶生；苞片漏斗状；花梗中上部有关节；花黄绿色，有短梗；花被5深裂，裂片矩圆形，长2～3mm；雄蕊7～8，比花被短；子房上位，1室，有3棱；花柱3，甚短，柱头头状。瘦果椭圆形，有3棱，黑色，平滑，有光泽。

〔生境〕

生于盐碱荒地或砂质含盐碱土壤。

〔用途〕

药用，具有清热解毒、利水渗湿功效。

〔Features〕

Perennial herb, rhizomes elongated. Stems obliquely ascending or nearly erect, 6～20cm tall, usually branched from the base. Leaves short petioles; leaf-blades oblong or lanceolate, nearly succulent, glabrous, 5～8cm long, 5～15mm wide, abruptly acute at the apex, hastate or cuneate at the base, the margin entire. Inflorescences coniform, terminal; bracts funnel-shaped; peduncles articulated in the upper middle section; yellowish green, short peduncles; perianth 5-parted, lobes oblong, about 2～3mm long; stamens 7～8, shorter than perianth; ovary superior, 1 locule, trigonous; stylus 3, very short, stigmas capitellate. Achenes oval, trigonous, black, smooth, lustrous.

〔Habitat〕

Saline-alkali wastelands or sandy saline alkali soil.

〔Use〕

Medicinal use, clearing heat and removing toxicity, clearing damp and promoting diuretic.

① 植株 ② 花 ③ 花蕾

荞麦属 *Fagopyrum* Mill.

苦荞麦 *Fagopyrum tataricum* (L.) Gaertn.

〔特征〕

一年生草本，高50～90cm。茎直立，分枝，绿色或略带紫色，有细条纹。叶有长柄；叶片宽三角形，长2～7cm，宽2.5～8cm，顶端急尖，基部心形，全缘；托叶鞘膜质，黄褐色。花序总状；花梗细长；花排列稀疏，白色或淡红色；花被5深裂，裂片椭圆形，长1.5～2mm，疏生细毛；雄蕊8，短于花被；花柱3，较短，柱头头状。瘦果卵形，有3棱，棱上部锐利，下部圆钝，黑褐色，长5～7mm，有3条深沟。

〔生境〕

生于村边、草地。

〔用途〕

饲用；药用；食用。

〔Features〕

Annual herb, 50～90cm tall. Stems erect, branches, green or slightly purply, with finely striate. Leaves long stems; leaf-blades broadly triangle, 2～7cm long, 2.5～8cm wide, acuminate at the apex, heart-shaped at the base, the margin entire; ocrea membranous, yellowish-brown. Inflorescence racemose; peduncles elongated; flowers sparsely arranged, white or light red; perianths 5-parted, lobes oval, about 1.5～2mm long, sparsely tiny hairs; stamens 8, shorter than perianths; stylus 3, short, stigma capitellate. Achene ovate, trigonous, sharp at the top, obtuse on the lower part, black brown, 5～7mm long, 3 deep grooves.

〔Habitat〕

Near villages and grasslands.

〔Use〕

Forage use; medicinal use; and edible use.

① 植株　② 叶

藜　科　Chenopodiaceae

藜　属　*Chenopodium* L.

藜　*Chenopodium album* L.

〔特征〕

一年生草本，高60～120cm。茎直立，粗壮，有棱和绿色或紫红色的条纹，多分枝；枝上升或开展。叶有长叶柄；叶片菱状卵形至披针形，长3～6cm，宽2.5～5cm，先端急尖中微钝，基部宽楔形，边缘常有不整齐的锯齿，下面生粉粒，灰绿色。花两性，数个集成团伞花簇，多数花簇排成腋生或顶生的圆锥状花序；花被片5，宽卵形或椭圆形，具纵隆脊和膜质的边缘，先端钝或微尖；雄蕊5；柱头2。胞果完全包于花被内或顶端稍露，果皮薄，和种子紧贴；种子横生，双凸镜形，直径1.2～1.5mm，光亮，表面有不明显的沟纹及点洼；胚环形。

〔生境〕

生于田间、路旁、荒地、宅旁等地。

〔用途〕

食用；药用，具有止泻痢，止痒功效。

〔Features〕

Annual herb, 60 ~ 120cm tall. Stems erect, stout, carinal and green or aubergine striped, multi-branched; branches ascending or squarrose. Leaves long petioles; leaf-blades rhombic-ovate to lanceolate, 3 ~ 6cm long, 2.5 ~ 5cm wide, acuminate to obtuse at the apex, broadly cuneate at the base, usually irregular serrate at the margin, particles beneath, grey green. Flowers bisexual, integrated into glomeruler, many flower culsters arranged into axillary or terminal conical inflorescence; tepals 5, broadly ovate or oval, longitudinal ridges and membranous at the margin, obtuse or acutate at the apex; stamens 5; stigmas 2. Utricles entirely wrapped in the perianths or apex slightly exposed, pericarp thin, seeds clingy; seeds amphitropous, lenticulate, 1.2 ~ 1.5mm in diameter, lustrous, inconspicuous fluted and pitted on the surface; embryo annular.

〔Habitat〕

Fields, roadsides, wastelands, and next to residential house, etc.

〔Use〕

Edible use; medicinal use, antidiarrheic and relieving itching.

① 植株 ② 叶、茎

猪毛菜属 *Salsola*

猪毛菜 *Salsola collina* Pall.

〔特征〕

一年生草本，高30 ~ 100cm。枝淡绿色，生稀疏的短糙硬毛或无毛。叶丝状圆柱形，肉质，生短糙硬毛，长2 ~ 5cm，宽0.5 ~ 1mm，先端有硬针刺。花序穗状，生枝条上部；苞片宽卵形，先端有硬针刺；小苞片2，狭披针形，比花被长；花被片5，膜质，披针形，长约2mm，果时背部生三翅或革质突起；花药矩圆形，顶部无附属物；柱头丝形，长为花柱的1.5 ~ 2倍。胞果倒卵形，果皮膜质。横生或斜生，直径1.5mm，顶端平；胚螺旋状；无胚乳。

〔生境〕

生于村边、路旁、荒地、含盐碱的沙质土壤上。

〔用途〕

饲用；药用，降血压。

〔Features〕

Annual herb, 30 ~ 100cm tall. Branches light green, sparse hispid bristles or glabrous. Leaves filiform cylindrical, succulent, hispid bristles, 2 ~ 5cm long, 0.5 ~ 1mm wide, hard stylose at the apex. Inflorescence spicate, branched at the top; bracts broadly ovate, hard stylose at the apex; bractlets 2, narrowly lanceolate, longer than perianths; tepals 5, membranous, lanceolate, about 2mm long, tripterous or leathery protruding on the dorsal surface during the fruiting period; anther oblong, no appendants on the top; stigma filate, about 1.5 ~ 2 times as long as stylus. Utricle obovate, pericarp membranous. Amphitropous or oblique, 1.5mm in diameter, flat at the apex; heliciform; no endosperm.

〔Habitat〕

Near villages, roadsides, wastelands and saline-alkali sandy soils.

〔Use〕

Forage use; medicinal use, antihypertensive effects.

① 植株
② 植株
③ 叶、茎
④ 花序
⑤ 叶
⑥ 花蕾

石竹科　Caryophyllaceae

蝇子草属　*Silene* L

旱麦瓶草　*Silene jenisseensis* Willd.

〔特征〕

多年生草本，高20～40cm。根圆柱形或纺锤形，淡黄色。茎丛生，直立，多分枝。叶对生，叶条形或条状披针形，顶端渐尖，基部成短鞘围抱节上，全缘或具细锯齿。聚伞花序顶生，叶腋分枝，每分枝上有2～3朵花；花萼圆筒形或棍棒形，长7～15mm，宽3～7mm，无毛，有10条纵脉，裂片钻状披针，绿色或紫绿色；花瓣5，白色或淡黄色，基部有长爪，瓣片顶端2深裂，喉部有2小鳞片；雄蕊10；子房上位，球形，花柱3。蒴果矩圆状卵形，种子多数，褐色。

〔生境〕

生于路旁、林下、草地及上坡草丛中。

〔用途〕

药用。

〔Features〕

Perennial herb, 20 ~ 40cm tall. Roots cylindrical or fusiform, faint yellow. Stems fasciculate, erect, multi-branched. Leaves opposite, leaves striped or striped lanceolate, acuminate at the apex, short sheath embracing internodes at the base, the margin entire or serrulate. Cymes terminal, axils branched, each branch 2 ~ 3-flowered; calyx cylindrical or clavate, 7 ~ 15mm long, 3 ~ 7mm wide, glabrous, 10 longitudinal veins, lobes subulate lanceolate, green or purple green; petals 5, white or light yellow, long claws at the base, limb 2-parted at the apex, 2 squamules in the throat; stamens 10; ovary superior, spherical, stylus 3. Capsule rounded-rectangular ovate, seeds numerous, brown.

〔Habitat〕

Roadsides, forests, grasslands and uphill bushes.

〔Use〕

Medicinal use.

① 植株 ② 花 ③ 叶 ④ 花蕾

石竹属 *Dianthus* L.

石 竹 *Dianthus chinensis* L.

〔特征〕

多年生草本，高约30cm。茎簇生，直立，无毛。叶条形或宽披针形，有时为舌形，长3～5cm，宽3～5mm。花顶生于分叉的枝端，单生或对生，有时成圆锥状聚伞花序；花下有4～6苞片；花萼圆筒形，萼齿5；花瓣5，鲜红色、白色或粉红色，花瓣扇状倒卵形，边缘有不整齐浅齿裂，喉部有深色斑纹和疏生须毛，基部具长爪；雄蕊10，子房矩圆形，花柱2，丝形。蒴果矩圆形；种子灰黑色，卵形，微扁，缘有狭翅。

〔生境〕

生于山地、田边和路旁，常见于山地草原。

〔用途〕

药用，利尿。

〔Features〕

Perennial herb, about 30cm tall. Stems fascicular, erect, glabrous. Leaves striped or broadly lanceolate, sometimes liguliform, 3 ~ 5cm long, 3 ~ 5mm wide. Flowers terminal on bifurcate ends of the branches; solitary or opposite, sometimes coniform cymes; 4 ~ 6 bracts; calyx cylindrical, calyx teeth 5; petals 5, bright red, white or pink, petals fan-shaped obovate, irregularly shallow dentate fissure at the margin, throat dark striped and sparse cirrus, long claws at the base; stamens 10, ovary oblong, stylus 2, filate. Capsule oblong; seeds grey black, ovate, slightly flat, with a narrow wing on the edge.

〔Habitat〕

Mountain lands, fields and roadsides, commonly seen on mountain lands and grasslands.

〔Use〕

Medicinal use, diuretic.

① 植株 ② 花蕾 ③ 花序 ④ 花蕾 ⑤ 花序 ⑥ 果实 ⑦ 景观

毛茛科 Ranunculaceae

翠雀属 *Delphinium* L.

翠 雀 *Delphinium grandiflorum* L.

〔特征〕

多年生草本，茎高35～70cm。全株有短毛。茎少分枝。基生叶和茎下部叶具长柄；叶片多圆肾形，长2.2～6cm，宽4～8cm，掌状3全裂，裂片细裂，小裂片条形，宽0.6～2.5mm，两面疏伏毛。总状花序具3～15花，轴和花梗被反曲的微柔毛；小苞片条形或钻形；萼片5，蓝色或紫蓝色，长1.5～1.8cm，距通常较萼片稍长，钻形，长1.7～2(2.3)cm；花瓣2，有距；退化雄蕊2，瓣片宽倒卵形，微凹，有黄色髯毛；雄蕊多数；心皮1～3(4)。蓇葖果1～3，有短毛。

〔生境〕

生于草甸草原、山地草坡、砾石质坡地、丘陵地带等。

〔用途〕

药用，治牙痛；有毒，茎叶浸汁可杀虫。

〔Features〕

Perennial herb, stems 35 ~ 70cm tall. The whole plant short-pilose. Stems rarely branched. Basal leaves and bottom leaves of stems have long petioles; leaf-blades mostly rounded-reniform, 2.2 ~ 6cm long, 4 ~ 8cm wide, palmate 3 pinnatisect, lobes finely fissus, small lobes striped, 0.6 ~ 2.5mm wide, sparse adpressed hairs on both sides. Racemes 3 ~ 15-flowered, rachis and peduncles retroflexed puberulent; bractlets striped or subulate; sepals 5, blue or purple blue, 1.5 ~ 1.8cm long, usually slightly longer than sepals, subulate, 1.7 ~ 2(2.3)cm long; petals 2, calcarate; staminodes 2, limb broadly obovate, slightly concave, yellow beard; stamens numerous; carpels 1 ~ 3(4). Follicles 1 ~ 3, short-pilose.

〔Habitat〕

Meadow steppes, mountain lands, grass slopes, sandy sloping fields, and hills, etc.

〔Use〕

Medicinal use for treating toothache; poisonous, stem leaves can be used for pesticidal.

① 植株 ② 花蕾

唐松草属 *Thalictrum* L.

展枝唐松草 *Thalictrum squarrosum* Steph. ex Willd.

〔特征〕

多年生草本，无毛。茎高60～100cm，常自中部分枝。茎下部及中部叶具短柄，近向上直展，为三至四回三出羽状复叶；叶片长8.5～18cm，小叶倒卵形，宽倒卵形或圆卵形，长0.8～2cm，宽0.6～1.5cm，通常2～4裂，裂片卵形或狭卵形，全缘或具2～3小牙齿，下面被白粉，脉平或稍隆起。圆锥花序稍呈伞房状，近二叉状分枝；花梗长1.5～3cm；花直径2.5～5mm；萼片4，淡黄绿色，狭卵形，长3mm；无花瓣；雄蕊多数，长3～5mm，花丝丝状；心皮1～3(5)，柱头三角形，具翅。瘦果近纺锤形或矩圆状倒卵形，无柄，伸直或稍弯曲，长4～6mm，纵肋8。

〔生境〕

生于平原草地、干燥草坡或田边、草甸草原、沙砾质草原等地。

〔用途〕

种子含液体油，供工业用。

〔Features〕

Perennial herb, glabrous. Stems 60～100cm tall, usually branched from the middle part. Stem leaves on lower and middle parts have short petioles, unfolding upward, tripinnate or quadripinnate pinnately compound leaves; leaf-blades 8.5～18cm,lobules obovate long, broadly obovate or rounded-ovate, 0.8～2cm long, 0.6～1.5cm wide, usually 2～4-lobed, lobes ovate or narrowly ovate, the margin entire or 2～3 denticles, pruinoses, veins plane or slightly ridgy. Panicles lightly corymbose, nearly dichotomous branching; peduncles 1.5～3cm long; flowers 2.5～5mm in diameter; sepals 4, chartreuse, narrowly ovate, 3mm long; no petals; stamens numerous, 3～5mm long, filaments filiform; carpels 1～3(5), stigmas triangle, with wings. Achenes nearly fusiform or rounded-rectangular obovate, impedicellate, straight or slightly curved, 4～6mm long, longitudinal ribs 8.

〔Habitat〕

Plains, grasslands, dry grass slopes or fields, meadow steppes and gravel grasslands, etc.

〔Use〕

Seeds containing liquid oil for industrial use.

① 植株 ② 花、花蕾 ③ 叶 ④ 花序

瓣蕊唐松草 *Thalictrum petaloideum* L.

〔特征〕

多年生草本，多分枝。叶为三至四回三出羽状复叶；小叶倒卵形、近圆形或菱形，长3～12mm，宽2～25mm，3浅裂至3深裂，裂片卵形倒卵形，全缘，脉平或微隆起。复单歧聚伞花序伞房状；花梗长5～28mm；花直径1～2cm；萼片4，白色，卵形，长3～5mm，先端圆，早落；无花瓣；雄蕊多数，长5～12mm，花丝中上部呈棍棒状，狭倒披针形，花药黄色，椭圆形；心皮4～13，无柄；花柱短，柱头狭椭圆形，稍外弯。瘦果无梗，卵状椭圆形，长4～6mm，宽2～3mm，先端尖，呈喙状，稍弯曲，具8条纵肋棱。花期6—7月，果期8月。

〔生境〕

生于草甸、草甸草原及山地沟谷中。

〔用途〕

药用，具有泻火解毒、消食、开胃功效。

〔Features〕

Perennial herb, multi-branched. Leaves tripinnate to quadripinnate ternate leaves compound leaves; lobules obovate, suborbicular or rhombic, 3～12mm long, 2～25mm wide, 3-lobed to 3-parted, lobes ovate obovate, the margin entire, veins plane or slightly ridgy. Compound monochasium corymbose; peduncles 5～28mm long; flowers 1～2cm in diameter; sepals 4, white, ovate, 3～5mm long, rounded at the apex, caducous; apetalous; stamens numerous, 5～12mm long, filaments clavate at the top, narrowly oblanceolate, anthers yellow, oval; carpels 4～13, impedicellate; stylus short, stigmas narrowly oval, slightly recurved. Achenes sessile, oval, 4～6mm long, about 2～3mm wide, acute at the apex, coronoid, slightly curved, 8 longitudinal ribs. Flowering June to July, and fruiting August.

〔Habitat〕

Meadows, meadow steppes, mountain lands and cheuches.

〔Use〕

Medicinal use, detoxicating, promoting digestion, and appetizing.

① 植株 ② 花序 ③ 花与叶 ④ 花序

铁线莲属 *Clematis* L.

棉团铁线莲 *Clematis hexapetala* Pall.

〔特征〕

多年生草本，高65～100cm，茎直立有纵棱，疏生短毛。叶对生，为一至二回羽状全裂；小叶革质，下部小叶不等2或3裂，裂片狭卵形至条形，中部小叶常2裂，上部小叶不分裂，披针形，全缘，网脉明显；叶柄长0.5～3.5cm。聚伞花序腋生或顶生，通常具3花；苞片条状披针形；花梗有伸展的柔毛；萼片6，白色，展开，狭倒卵形，长1.5～1.7cm，宽6～10mm，顶端圆形，外面有白色绵毛，无花瓣，雄蕊多数，长药9mm，无毛；心皮多数。瘦果倒卵形，扁平，长约4mm，有紧贴的柔毛，羽毛状花柱长达2.2cm。

〔生境〕

生于山地、干燥山坡或草坡上。

〔用途〕

药用，具有解热、镇痛、利尿、痛经功效。

〔Features〕

Perennial herb, 65～100cm tall, stems erect with longitudinal ridges, sparsely pubescent. Leaves opposite, unipinnate to bipinnatisect; lobules leathery, lower part lobules 2 or 3-lobed, lobes narrowly ovate to striped, central lobules bifid, lobules non-splitting at the top, lanceolate, the margin entire, reticulate veins conspicuous; petioles 0.5～3.5cm long. Cymes axillary or terminal, usually 3-flowered; bracts striped-lanceolate; peduncles have stretched pubescence; sepals 6, white, unfolding, narrowly obovate, 1.5～1.7cm long, 6～10mm wide, rounded at the apex, white-lanate outside, apetalous, stamens numerous, anther 9mm long, glabrous; carpel numerous. Achenes obovate, flat, about 4mm long, pubescence appressed, penniform stylus 2.2cm long.

〔Habitat〕

Mountain lands, dry hillsides or grass slopes.

〔Use〕

Medicinal use, antipyretic, analgesic, diuretic, and relieving the symptoms of dysmenorrhea.

① 植株 ② 花 ③ 景观 ④ 花序

白头翁属 *Pulsatilla* Adans

白头翁 *Pulsatilla chinensis*（Bunge）Regel，Tent.

〔特征〕

植株高15 ~ 35cm。根状茎粗0.8 ~ 1.5cm。基生叶4 ~ 5，通常在开花时刚刚生出，有长柄；叶片宽卵形，长4.5 ~ 14cm，宽6.5 ~ 16cm，三全裂，中全裂片有柄或近无柄，宽卵形，三深裂，中深裂片楔状倒卵形，少有狭楔形或倒梯形，全缘或有齿，侧深裂片不等二浅裂，侧全裂片无柄或近无柄，不等三深裂，表面变无毛，背面有长柔毛；叶柄长7 ~ 15cm，有密长柔毛。花葶1（2），有柔毛；苞片3，基部合生成长3 ~ 10mm的筒，三深裂，深裂片线形，不分裂或上部三浅裂，背面密被长柔毛；花梗长2.5 ~ 5.5cm，结果时长达23cm；花直立；萼片蓝紫色，长圆状卵形，长2.8 ~ 4.4cm，宽0.9 ~ 2cm，背面有密柔毛；雄蕊长约为萼片之半。聚合果直径9 ~ 12cm；瘦果纺锤形，扁，长3.5 ~ 4mm，有长柔毛。宿存花柱长3.5 ~ 6.5cm，有向上斜展的长柔毛。4—5月开花。

〔生境〕

生于平原和低山山坡草丛中、林边或干旱多石的坡地。

〔用途〕

药用；根状茎水浸液亦可作土农药。

〔Features〕

The plant 15 ~ 35cm tall. Rhizomes 0.8 ~ 1.5cm thick. Basal leaves 4 ~ 5, usually grow when blooming, with long petiole; leaf-blades broadly ovoid, 4.5 ~ 14cm long, 6.5 ~ 16cm wide, pinnatisect, central lobes petiolate or nearly impedicellate, broadly ovoid, 3-parted, central lobes sphenoid obovate, rarely narrowly sphenoid or inverted trapezoidal, the margin entire or dentate, lateral lobes 2-lobed, lateral lobes impedicellate or nearly impedicellate, 3-parted, glabrous on the surface, long pubescent on the back; petioles 7 ~ 15cm long, densely long pubescent. Scape 1 or 2, pubescent; bracts 3, concrescent to a 3 ~ 10mm-long tube at the base, 3-parted, deep lobes linear, not divided or 3-lobed at upper part, densely long pubescent on the back; pedicels 2.5 ~ 5.5cm long, up to 23cm long when fruiting; flowers erect; sepals bluish violet, oblong-elliptic ovoid, 2.8 ~ 4.4cm long, 0.9 ~ 2cm wide, densely pubescent on the back; stamens about half of the sepals length. Aggregated fruits 9 ~ 12cm in diameter; achenes spindle-shaped, flat, 3.5 ~ 4mm long, long pubescent. Persistent stylus 3.5 ~ 6.5cm long, upward obliquely spreading long pubescent. Flowering April to May.

〔Habitat〕

Plains, low hillside, bushes, forest edges or dry stony sloping fields.

〔Use〕

Medicinal use; rhizomes infusion can also be used as native pesticides.

① 植株 ② 花、花蕾 ③ 茎、叶 ④ 小花

十字花科 Cruciferae

花旗竿属 *Dontostemon* Andrz.ex Ledeb.

小花花旗竿 *Dontostemon micranthus* C. A. Mey.

〔特征〕

一年生或二年生草本，植株高15～45cm，被直立糙毛及柔毛。茎单一或分枝，少数具莲座状基生叶；茎生叶较密集着生，线形，长1.5～4cm，宽2～5mm，全缘，两面略具毛，边缘具糙毛。总状花序顶生，结果时长12～25cm，花小；萼片线状披针形，长约2.5mm，宽约0.8mm，具白色膜质边缘，背部稍具糙毛；花瓣淡紫色或白色，线状长椭圆形，长3.5～5mm，为萼片长度的1.5倍，宽约1mm，顶端钝，基部渐狭成短爪。长角果径直或弯曲，近圆柱形，长2～3.5cm，光滑无毛；花柱极短，柱头稍增大，果梗斜上开展。种子褐色，椭圆形，细小，无膜质边缘；子叶背倚（稀为斜缘倚）胚根。花果期6—8月。

〔生境〕

生于山坡草地、河滩、固定砂丘及山沟。

〔用途〕

药用。

〔Features〕

Annual or biennial herb, plant 15～45cm tall, erect hispid and pubescent. Stems single or branched, a few with rosette basal leaves; cauline leaves densely inserted, linear, 1.5～4cm long, 2～5mm wide, the margin entire, slight hairs on both surfaces, hispid at the margin. Racemes terminal, 12～25cm long during the fruiting period, flowers small; sepals linear lanceolate, about 2.5mm long, about 0.8mm wide, with white membranous margins, slightly strigose on the dorsal surface; petals lavender or white, linear oblong, 3.5～5mm long, about 1.5 times as long as sepals, about 1mm wide, obtuse at the apex, tapering into short claws at the base. Silique straight or curved, nearly cylindrical, 2～3.5cm long, glabrous; stylus very short, stigma slightly accrescent, pedicels obliquely ascending. Seeds brown, oval, tiny, no membranous margins; Cotyledons leaning against (rarely recumbent) radicles. Flowering and fruiting June to August.

〔Habitat〕

Hillsides, grasslands, beach lands, fixed sand dunes and valleys.

〔Use〕

Medicinal use.

① 植株 ② 花蕾 ③ 叶

独行菜属 *Lepidium* L.

独行菜 *Lepidium apetalum* Willd.

〔特征〕

一年生或二年生草本。茎高20～30cm，上部多分枝并有乳头状短毛。叶互生，无柄，基部有耳状小裂片；茎下部叶狭矩圆形或狭匙形，长3～5cm，宽1～1.5cm，羽状浅裂或深裂并具1～2cm柄；茎上部叶条形，边缘有疏齿或全缘。总状花序顶生；花极小；萼片4枚，开花后早落，边缘白色，但通常先端紫褐色；花瓣丝状4枚，退化；雄蕊2～4，具4蜜腺点；子房一个。短角果近圆形或椭圆形，扁平，长2.5～3mm，宽2mm，先端微凹，上部具窄翅；种子椭圆形，棕红色，长约1mm，光亮。

〔生境〕

生于路旁、沟边、宅旁。

〔用途〕

种子入药，有利尿、止咳化痰之效。

〔Features〕

Annual or biennial herb. Stems 20～30cm tall, multi-branched at the top and capitellate short-pilose. Leaves alternate, impedicellate, auriform lobelets at the base; leaves narrowly oblong or narrowly spathulate at lower part of stems, 3～5cm long, 1～1.5cm wide, pinnate lobed or parted with 1～2cm petioles; leaves striped at the top of stems, sparse denticulate at the margin or the margin entire. Racemes terminal; flower tiny; sepals 4, caducous after flowering, white at the margin, but usually purple brown at the apex; 4-petals filiform, vestigial; stamens 2～4, 4nectary glands; 1 ovary. Silicle suborbicular or oval, flat, 2.5～3mm long, 2mm wide, concave at the apex, narrow winged at the top; seeds oval, brown red, about 1mm long, lustrous.

〔Habitat〕

Roadsides, ditches and next to residential houses.

〔Use〕

Seeds used as medicine, diuretic, relieving cough and reducing sputum.

① 植株 ② 叶

播娘蒿属 *Descurainia* Webb et Berth.

播娘蒿 *Descurainia sophia* (L.) Webb. ex Prantl

〔特征〕

一年生草本，高30 ~ 70cm，密被灰白色分枝毛或星状毛。茎上部分枝并灰白色。叶二回至三回羽状分裂或深裂，长3 ~ 5cm，宽2 ~ 2.5cm；末回小裂片细条形，长3 ~ 5mm，宽1 ~ 1.5mm；茎下部叶有柄，茎上部叶无柄。总状花序顶生，小花多数，直径2mm；萼片4枚，早落；花瓣4枚，淡黄色，长2 ~ 2.5mm；雄蕊6；雌蕊圆柱状并柱头扁穗形。长角果条形，稍有节，长2.5 ~ 3cm，无毛，有1 ~ 2cm长柄；种子小，多数，每室有一行，长圆形，褐色。

〔生境〕

生于草原、山坡草地，潮湿地生长更旺盛。

〔用途〕

种子入药；全草可作农药。

〔Features〕

Annual herb, 30 ~ 70cm tall, densely covered with hoary branched hairs or stellate hairs. Branches at the top of stems hoary. Leaves bipinnately to tripinnately lobed or parted, 3 ~ 5cm long, 2 ~ 2.5cm wide; lobelets finely striped, 3 ~ 5mm long, 1 ~ 1.5mm wide; leaves at lower part of stems petiolate, leaves at the top of stems impedicellate. Racemes terminal, florets numerous, 2mm in diameter; sepals 4, caducous; petals 4, light yellow, 2 ~ 2.5mm long; stamens 6; pistils cylindrical and stigmas flat spike-shaped. Silique striped, slightly jointed, 2.5 ~ 3cm long, glabrous, 1 ~ 2cm long petioles; seeds small, numerous, each locule has a line, long elliptic, brown.

〔Habitat〕

Grasslands, hillsides grasslands, and more vigorous at wet wetland.

〔Use〕

Seeds can be used as medicine; the whole herb can be used as pesticide.

① 植株
② 顶拍
③ 顶拍

景天科 Crassulaceae

瓦松属 *Orostachys* Fisch.

瓦　松 *Orostachys fimbriatus*(Turcz.)Berger

〔特征〕

二年生肉质草本，高10～30cm，全株粉绿色，密生紫红色斑点；莲座叶和茎生叶先端具刺头；花序总状或圆锥状；花瓣5，粉红色，先端具凸尖头；花药紫色；蓇葖5，长圆形，长5mm，喙细，长1mm；种子多数，卵形，细小。花果期8—9月。

〔生境〕

生于石质山坡、石质丘陵及沙质地。

〔用途〕

全草入药。有止血、活血、敛疮之效。

〔Features〕

Biennial succulent herbal, 10～30cm tall, the whole plant pastel green, densely aubergine spots; rosette leaves and cauline leaves aculeate at the apex; Inflorescence racemose or coniform; petals 5, pink, sharply cuspidate at the apex; anthers purple; follicles 5,long elliptic, 5mm long, rostrum slender, 1mm long; seeds numerous, ovoid, tiny. Flowering and fruiting August to September.

〔Habitat〕

Stony hillside, stony hills and sandy land.

〔Use〕

The whole herb used as medicine. Hemostatic, activating blood circulation and healing sore.

① 植株 ② 花蕾 ③ 花 ④ 叶

景天属 *Sedum* L.

费 菜 *Sedum aizoon* L.

〔特征〕

多年生草本，肉质，茎直立，高20～50cm。根状茎肥大而稍木质化。叶互生或对生，肉质，宽披针形至细倒楔形，长3～8cm，宽1.7～2cm，基部楔形。边缘有不规则的粗锯齿，叶下部全缘，光滑或稍有乳头状腺点。花无梗或有极短柄，伞房状聚伞花序；萼片不等长条状5裂，长3～5mm；花瓣5枚，黄色，长6～10mm；雄蕊10，比花瓣短；心皮5枚，卵状长圆形，基部合生。蓇葖果星角状排列；种子光滑并边缘具翅。

〔生境〕

生于山坡潮湿草地、岩石上。

〔用途〕

药用。具有止血、消肿止痛功效。

〔Features〕

Perennial herb, succulent, stems erect, 20～50cm tall. Rhizomes hypertrophic and slightly lignified. Leaves alternate or opposite, succulent, broadly lanceolate to obcuneate, 3～8cm long, 1.7～2cm wide, cuneate at the base. Irregularly serrate at the margin, the margin entire at lower part of leaves, glabrous or slightly capitellate glandular dots. Flowers sessile or very short petioles, corymbose cymes; sepals unequal length striped 5-lobed, 3～5mm long; petals 5, yellow, 6～10mm long; stamens 10, shorter than petals; carpels 5, oval long elliptic, concrescent at the base. Follicles cornute arranged; seeds glabrous and epipterous at the margin.

〔Habitat〕

Hillsides, wet grasslands and rocks.

〔Use〕

Medicinal use, arresting bleeding, detumescent and pain relieving.

① 植株及横走茎 ② 株丛 ③ 叶 ④ 地上横走茎

蔷薇科 Rosaceae

委陵菜属 *Potentilla* L.

绢毛细蔓委陵菜 *Potentilla reptans* L. var.*sericophylla* Franch.

〔特征〕

叶为三出掌状复叶，边缘两个小叶浅裂至深裂，有时混生有不裂者，小叶下面及叶柄伏生绢状柔毛，稀脱落被稀疏柔毛。花果期4—9月。

〔生境〕

生于山坡草地、渠旁、溪边灌丛中及林缘。

〔用途〕

药用，具收敛解毒、生津止渴、止咳功效。

〔Features〕

Leaves ternately palmate leaves, 2 leaflets lobed to parted at the margin, occasionally indehiscent, silk pubescent below leaflets and petioles, rarely deciduous sparse pubescent. Flowering and fruiting April to September.

〔Habitat〕

Hillside meadows, channels, bushwoods by streams and forest edges.

〔Use〕

Medicinal use, antidotic, encouraging production of body fluids to extinguish thirst, and cough relieving.

① 植株 ② 花 ③ 叶 ④ 景观

星毛委陵菜 *Potentilla acaulis* L.

〔特征〕

多年生草本，茎短，高2～8cm，全株被星状毡毛。三出复叶；小叶菱状倒卵形或倒卵状楔形，长8～15mm，宽6～8mm，顶端钝、基部楔形，边缘钝锯齿，但叶中部以下全缘，两面密被星状毛；托叶披针形，被毛；叶柄长0.3～1cm，密被星状毛或长柔毛。花黄色，直径1.5cm，常单生，或2～3朵集成聚伞花序，花梗长约1.5cm；萼片5枚，短披针形，外边有卵形副萼片，都被星状毛或毡毛；花瓣5枚，基部宽楔形；雌雄蕊都多数；雌蕊生在圆锥状花托上。瘦果长卵形，褐色，多褶皱。

〔生境〕

生于草原、草甸草原、山地草坡。

〔用途〕

饲用；亦有保持水土作用。

〔Features〕

Perennial herb, stems short, 2～8cm tall, and the whole plant covered with starlike pannose. Ternately compound leaves; lobules rhombic obovate or obovate cuneate, 8～15mm long, 6～8mm wide, obtuse at the apex, cuneate at the base, obtuse serrulate at the margin, but the margin entire at the middle part and the below, densely covered with stellate hairs on both sides; stipules lanceolate, pilose; petioles 0.3～1cm long, densely covered with stellate hairs or long pubescence. Flowers yellow, 1.5cm in diameter, usually solitary, or 2～3 flowers clustered into cymes, peduncles about 1.5cm long; sepals 5, short lanceolate, ovate epicalyx segments outside, clothed with stellate hairs or pannose; petals 5, broadly cuneate at the base; both pistils and stamens numerous; pistils born on the coniform receptacles. Achenes long ovate, brown, plicated.

〔Habitat〕

Grasslands, meadow steppes, mountain lands and grass slopes.

〔Use〕

Forage use; soil and water conservation.

① 植株
② 花
③ 叶
④ 花与花蕾
⑤ 花蕾

菊叶委陵菜 *Potentilla tanacetifolia* Willd.

〔特征〕

多年生草本，高15～50cm，根木质化。基部有许多老叶的残留物；茎直立或斜生；茎、叶柄、花序全被长毛或短毛。基生叶单数羽状复叶，叶柄长，有小叶9～15；小叶倒卵状椭圆形或椭圆形，长1～3cm，宽0.6～1.2cm，边缘具缺刻状锯齿或浅裂，上面几乎无毛，背面被柔毛，无柄；上部小叶较大；茎生叶2～5对，有短柄；托叶椭圆形，具披针形两个裂片。伞房状聚伞花序较大，花黄色，多数，直径8～15mm；萼片5枚，外面有柔毛和腺点，副萼片5枚；花瓣5枚，雌雄蕊多数。瘦果长卵形，淡黄绿色，光滑，上部有宿存花柱。花果期7—10月。

〔生境〕

生于草原、草甸草原、向阳坡地上。

〔用途〕

药用，可治痢疾、结核、吐血等症。

〔Features〕

Perennial herb, 15～50cm tall, roots lignified. Many residues of old leaves at the base; stems erect or oblique; stems, petioles and inflorescences long pubescent or short-pilose. Odd basal leaves pinnately compound leaves, petioles long, 9～15 lobules; lobules obovate oval or oval, 1～3cm long, 0.6～1.2cm wide, erose serrulate or lobed at the margin, nearly glabrous at the top, pubescent on the back, impedicellate; lobules bigger at the top; cauline leaves 2～5 pairs, short petioles; stipules oval, with lanceolate 2 lobes. Corymbose cymes larger, flowers yellow, numerous, 8～15mm in diameter; sepals 5, pubescence and glandular dots on the outside, epicalyx segments 5; petals 5, pistils and stamens numerous. Achene long ovate, chartreuse, glabrous, persistent stylus at the top. Flowering and fruiting July to October.

〔Habitat〕

Grasslands, meadow steppes, sunny sloping fields.

〔Use〕

Medicinal use, treating dysentery, tuberculosis and hematemesis, etc.

① 植株 ② 花、茎、叶 ③ 叶背面

三出叶委陵菜 *Potentilla betonicaefolia* Poir.

〔特征〕

多年生草本，高10 ~ 20cm。茎直立或斜升，无匍匐分枝。基生叶掌状三出复叶，具三小叶，稀五小叶；小叶片矩圆状披针形，鳞片状，顶端渐尖，基部圆形或楔形，边缘具粗锯齿，上面无毛，有光泽，绿色，背面密被白色毡毛；叶柄较长，无毛；茎生叶无柄，极小。聚伞花序顶生，具总花序梗；叶柄无绒毛；萼片5枚，副萼片5枚，全被白毛；花瓣5枚，黄色，近圆形或倒卵形，雄蕊、雌蕊都多数。瘦果多数，多毛的圆锥状花托上密生；花托熟后枯萎。

〔生境〕

生于草原、丘陵草地、山地砾石坡上。

〔用途〕

药用。

〔Features〕

Perennial herb, 10 ~ 20cm tall. Stems erect or obliquely ascending, no decumbent branches. Basal leaves palmate ternately compound leaves, trifoliolate, rarely quinate; small leaf-blades rounded-rectangular lanceolate, scaly, tapering at the apex, rounded or cuneate at the base, serrate at the margin, glabrous at the top, lustrous, green, densely white pannose on the back; petioles long, glabrous; cauline leaves impedicellate, tiny. Cymes terminal, inflorescence peduncles; petioles laeve; sepals 5, epicalyx segments 5, covered with white hairs; petals 5, yellow, suborbicular or obovate, stamens and pistil numerous. Achenes numerous, dense on the hairy coniform receptacles; receptacles withered when mature.

〔Habitat〕

Grasslands, hilly grasslands, mountain lands and gravel slopes.

〔Use〕

Medicinal use.

① 植株 ② 花 ③ 叶 ④ 花 ⑤ 叶

二裂叶委陵菜 *Potentilla bifurca* L.

〔特征〕

矮小多年生草本，高10 ~ 20cm，根木质化，棕褐色，茎秆细，基部多分枝，匍匐地上或斜升，被长毛。基生叶羽状复叶，小叶5 ~ 17；小叶条状披针形或椭圆形，或倒卵状椭圆形，长0.5 ~ 1.5cm，顶端锐尖，基部楔形，全缘，上面有毛，背面被柔毛；叶柄长2 ~ 8mm，常被柔毛；托叶细长披针形。基部与叶柄合生；茎生小叶3 ~ 7个。花单生或3 ~ 5朵集成聚伞花序，直茎1 ~ 1.5cm，花梗长1 ~ 2.5cm，被毛；花瓣5枚，黄色。雄蕊多数，中间生绢毛；雌蕊多数，每个雌蕊有下垂生的胚珠，密被柔毛的花托上簇生。瘦果小，光滑，花柱宿存。花果期5—8月。

〔生境〕

生于草原、山坡、田边、路旁。

〔用途〕

药用，具有清热解毒、收敛止血功效。

〔Features〕

Stunted perennial herb, 10 ~ 20cm tall, roots lignified, brown, stems slender, multi-branched at the base, decumbent or obliquely ascending, long pubescent. Basal leaves pinnately compound leaves ,lobules 5 ~ 17; lobules striped lanceolate or oval, or obovate-oval, 0.5 ~ 1.5cm long, acuminate at the apex, cuneate at the base, the margin entire, hairy at the top, pubescent on the back; petioles 2 ~ 8mm long, constantly pubescent; stipules slender, lanceolate. Petioles concrescent at the base; cauline lobules 3 ~ 7. Flowers solitary or 3 ~ 5 flowers clustered into cymes, 1 ~ 1.5cm in diameter, peduncles 1 ~ 2.5cm long, pilose; petals 5, yellow. Stamens numerous, sericeous in the center; pistil numerous, each has pendulous ovules, densely pubescent receptacles fascicular. Achenes small, glabrous, stylus persistent. Flowering and fruiting May to August.

〔Habitat〕

Grasslands, hillsides, fields and roadsides.

〔Use〕

Medicinal use, clearing heat and removing toxicity and anastaltic.

① 植株 ②花与花蕾 ③ 叶

高二裂委陵菜 *Potentilla bifurca* L. var. *maior* Ldb.

〔特征〕

多年生草本或亚灌木。根圆柱形，纤细，木质。茎直立或上升，高5～20cm，密被疏柔毛或微硬毛。羽状复叶，有小叶5～8对，最上面2～3对小叶基部下延与叶轴汇合，连叶柄长3～8cm；叶柄密被疏柔毛或微硬毛；小叶片无柄，对生稀互生，椭圆形或倒卵椭圆形，长0.5～1.5cm，宽0.4～0.8cm，顶端常2裂，稀3裂，基部楔形或宽楔形；两面绿色，伏生疏柔毛；下部叶托叶膜质，褐色，外面被微硬毛，稀脱落几无毛，上部茎生叶托叶草质，绿色，卵状椭圆形，顶端急尖，常全缘稀有齿。近伞房状聚伞花序，顶生，疏散；花直径0.7～1cm；萼片椭圆形，顶端急尖，副萼片椭圆形，顶端急尖或钝，比萼片短或近等长，外面被疏柔毛；花瓣黄色，倒卵形，顶端圆钝，比萼片稍长；心皮沿腹部有稀疏柔毛；花柱侧生，棒形，基部较细，顶端溢缩，柱头扩大。瘦果表面光滑。花果期5—9月。

〔生境〕

生于耕地道旁、河滩沙地、山坡草地。

〔用途〕

药用；饲用。

〔Features〕

Perennial herb or sub-shrub. Roots cylindrical, slender, woody. Stems erect or ascending, 5 ~ 20cm tall, sparse pubescent or hirtellous. Pinnately compound leaves, lobules 5 ~ 8 pairs, 2 ~ 3-paired lobules topside decurrent with rachises at the base, petioles 3 ~ 8cm long; petioles sparse pubescent or hirtellous; small leaf-blades impedicellate, opposite sparsely alternate, oval or obovate-oval, 0.5 ~ 1.5cm long, 0.4 ~ 0.8cm wide, usually bifid at the apex, rarely 3-lobed, cuneate or broadly cuneate at the base; green on both sides, sparse pubescent; bottom stipules membranous, brown, hirtellous outside, deciduous glabrescent, cauline leaves stipules herbaceous at the top, green, oval, abruptly acuminate at the apex, usually the margin entire, rarely denticulate. Nearly corymbose cymes, terminal, scattered; flowers 0.7 ~ 1cm in diameter; sepals oval, abruptly acuminate at the apex, epicalyx segments oval, abruptly acuminate or obtuse at the apex, shorter than sepals or approximately equal in length, sparse pubescent outside; petals yellow, obovate, obtuse at the apex, slightly longer than sepals; carpels sparse pubescent along the abdomen; stylus coadnate, claviform, slender at the base, shrunken at the apex, stigmas expanded. Achenes glabrous on the surface. Flowering and fruiting May to September.

〔Habitat〕

Roadsides, beach lands, sand lands, hillsides and grasslands.

〔Use〕

Medicinal use; forage use.

① 植株 ② 小花 ③ 茎叶

地蔷薇属 *Chamaerhodos* Bge.

地蔷薇 *Chamaerhodos erecta* (L.) Bge.

〔特征〕

二年生草本，根状茎木质化。茎纤细，直立，上部多分枝，通常高10～60cm，全株密被长柔毛和腺毛。叶互生，基部叶两深裂，小裂片再次2～4分裂，最末小裂片分裂成丝状；叶柄长1～2.5cm，茎生叶与基部叶基本相同；托叶膜质，两深裂。花小，有0.5～1cm长柄，直径2～3mm，集成圆锥状花序；花萼钟状，先端5裂，裂片披针形，宿存；小苞片通常两裂；花瓣5枚，粉红色或白色，倒卵形并有爪；雄蕊5，花丝基部膨大，于花瓣对生；花托边缘具刺毛；心皮10～15，密被毛；胚珠单生。瘦果卵形，黑色，无毛。

〔生境〕

生于山坡草地和沙质地。

〔用途〕

药用。

〔Features〕

Biennial herb, rhizomes lignified. Stems slender, erect, multi-branched at the top, usually 10～60cm tall, the whole plant long has pubescent and glandular hairs. Leaves alternate, basal leaves 2-parted, lobelets 2～4-lobed, terminal lobelets split into filiform; petioles 1～2.5cm long, cauline leaves basically the same as basal leaves; stipules membranous, 2-parted. Flowers small, 0.5～1cm long petioles, 2～3mm in diameter, clustered into conical inflorescence; calyx campaniform, 5-lobed at the apex, lobes lanceolate, persistent; bractlets usually bifid; petals 5, pink or white, obovate and unguiculate; stamens 5, filaments inflated at the base, petals opposite; receptacles have bristles at the margin; carpels 10～15, covered with dense hairs; ovules solitary. Achene ovate, black, glabrous.

〔Habitat〕

Hillsides, grasslands and sandy lands.

〔Use〕

Medicinal use.

① 植株 ② 花 ③ 茎、叶 ④ 扁茎秆

毛地蔷薇 *Chamaerhodos canescens* Krause

〔特征〕

多年生草本；根木质；茎多数，丛生，直立或上升，高10～30cm，上部分枝，基部密生短腺毛及疏生长柔毛。基生叶密集，长1～1.5cm，有腺毛及灰色长刚毛，二回三裂，一回裂片三深裂，二回裂片全缘或深缺刻状二至三裂，小裂片条形，长4～6mm，先端近圆钝或锐尖，基部楔形，全缘；茎生叶似基生叶，侧裂片常全缘，少有具缺刻，中裂片三深裂，二回裂片再二至三裂；基生叶的叶柄长1.5～3cm，茎生叶者极短，约长5mm，皆有长刚毛；托叶和茎生叶侧裂片相似，条形，约长5mm，全缘，有长刚毛。花成复聚伞花序，直径2～3cm，多花，排列紧密；总花梗及花梗有具腺柔毛；苞片及小苞片披针形，长5～10mm，二至三深裂，裂片条形，有具腺柔毛；花直径3～5mm；

花梗长2～4mm；萼筒宽钟形，长2～3mm，外面有长刚毛，萼片披针形，长约2mm，先端渐尖，有10条显明脉和长刚毛；花瓣倒卵形，长3～4mm，粉红色或白色，先端微缺，基部具短爪，无毛；花丝长1.5～2mm，无毛；花托有长柔毛；心皮4～6，离生，花柱丝状，子房无毛。瘦果长圆卵形，长2mm，黑褐色，无毛，先端渐尖具尖头。花期6—8月，果期8—10月。

〔生境〕

生于山坡岩石间。

〔用途〕

饲用。

〔Features〕

Perennial herb; roots woody; stems numerous, fasciculate, erect or ascending, 10～30cm tall, branches at the top, densely short glandular hairs and sparsely pubescent at the base. Basal leaves dense, 1～1.5cm long, glandular hairs and grey long bristles, bipinnately trilobite, lobes pinnately 3-parted, bipinnate lobes the margin entire or erose bilobate to trilobate, lobelets striped, 4～6mm long, obtuse or acute at the apex, cuneate at the base, the margin entire; cauline leaves similar to basal leaves, usually the margin entire of lateral lobes, a few erose, central lobes 3-parted, bipinnate lobes bilobate to trilobate; petioles of basal leaves 1.5～3cm long, cauline leaves very short, about 5mm long, long bristles; lateral lobes of stipules and cauline leaves are similar, striped, about 5mm long, the margin entire, with long bristles. Flowers compound cyme, 2～3cm in diameter, floriferous, closely arranged; peduncles and pedicels glandular pubescent; bracts and bractlets lanceolate, 5～10mm long, 2-parted to 3-parted, lobes striped, glandular pubescent; flowers 3～5mm in diameter; pedicels 2～4mm long; calyx tube broadly campaniform, 2～3mm long, long bristles on the outside, sepals lanceolate, about 2mm long, acuminate at the apex, 10 conspicuous veins and long bristles; petals obovate, 3～4mm long, pink or white, emarginated at the apex, short claws at the base, glabrous; filaments 1.5～2mm long, glabrous; receptacles long pubescent; carpels 4～6, dialypetalous, stylus filiform, ovary glabrous. Achenes long ovoid, 2mm long, dark brown, glabrous, acuminate and sharply cuspidate at the apex. Flowering June to August, and fruiting August to October.

〔Habitat〕

Hillside and between rocks.

〔Use〕

Forage use.

① 植株 ② 花序 ③ 株丛

地榆属 *Sanguisorba* L.

地榆 *Sanguisorba officinalis* L.

〔特征〕

多年生草本，高1～2m。全株无毛，根肉质。茎直立，具沟槽和棱。奇数羽状复叶互生，小叶3～15枚；基生叶具长柄；茎生叶几乎无柄，基部两边有抱茎的凹状托叶；小叶长椭圆形或矩圆形，长1.5～6cm，宽0.8～3cm，顶端尖或钝，基部心形或截形，边缘具锯齿，两边无毛；小叶柄短，基部有一对托叶具裂齿。花小，夏秋开花，花密集成长圆形穗状花序，花序柄细长，稍有毛；苞片膜质，条形或披针形并被毛；萼片4枚，暗红紫色，基部有毛；无花瓣；雄蕊4；子房有毛，花柱4个，比雄蕊短。瘦果球形，有纵沟，包裹在宿存花萼内。

〔生境〕

生于山地草原、草甸草原、村边及潮湿地上。

〔用途〕

药用。

〔Features〕

Perennial herb, 1～2m tall. The whole plant glabrous, roots succulent. Stems erect, fluted and carinal. Odd pinnately compound leaves alternate, lobules 3～15; basal leaves long petioles; cauline leaves nearly impedicellate, amplexicaul concave stipules on both sides of the base; lobules long oval or oblong, 1.5～6cm long, 0.8～3cm wide, acute or obtuse at the apex, heart-shaped or truncate at the base, serrulate at the margin, glabrous on both sides; petioles short, a pair of stipules denticulate at the base. Flowers small, blooming in summer and autumn, flowers clustered into long elliptic spike, inflorescence stems long and thin, slightly hairy; bracts membranous, strip or lanceolate and pilose; sepals 4, dark purple, hairy at the base; apetalous; stamens 4; ovary hairy, stylus 4, shorter than stamens. Achene spherical, with longitudinal furrows, enclosed in the persistent calyx.

〔Habitat〕

Mountain lands, grasslands, meadow steppes, villages and wetlands.

〔Use〕

Medicinal use.

① 植株 ② 小穗和叶

豆 科 Leguminosae

胡枝子属 *Lespedeza* Michx.

达乌里胡枝子 *Lespedeza davurica* (Laxm.) Schindl.

〔特征〕

通常稍斜升的较小灌木，高20～100cm，被白色绒毛。羽状三出复叶，互生；小叶披针状矩圆形或狭矩圆形，长2～3cm，宽7～10mm，先端钝，中间有短刺，全缘，背面密生短毛；托叶2枚，条形。总状花序腋生，比叶短；花柄无节；无花瓣的花在下部分枝叶腋内簇生，小苞片条形；花萼浅杯形，裂片5枚，顶端芒状，披针形，有白毛并与花冠近等长；花冠白色或淡黄绿色，旗瓣条状矩圆形，长约1cm，翼瓣较短，龙骨瓣比翼瓣长；子房无毛。荚果心形或倒卵状长圆形，长约4mm，宽2.5mm，有白色柔毛。花果期7—10月。

〔生境〕

生于平原草地、河滩、路旁以及山顶、山沟等。

〔用途〕

饲用；药用，治发热、咳嗽。

〔Features〕

Small shrub usually slightly oblique ascending, 20～100cm tall, covered with white tomentum. Pinnate ternately compound leaves, alternate; lobules lanceolate oblong or narrowly oblong, 2～3cm long, 7～10mm wide, obtuse at the apex, short thrust in the middle, the margin entire, dense short-pilose on the back; stipules 2, stripeded. Racemes axillary, shorter than leaves; pedicels enodal; apetalous flowers fascicular at lower part of axil, bractlets striped; calyx cyathiform, lobes 5, apex awned, lanceolate, with white hairs and approximately equal to corolla in length; corolla white or chartreuse, vexil oblong, about 1cm long, alae shorter, carinas longer than alae; ovary glabrous. Legumes heart-shaped or obovate-elliptic, about 4mm long, 2.5mm wide, with white pubescence. Flowering and fruiting July to October.

〔Habitat〕

Plain grasslands, beachlands, roadsides, mountaintops and valleys.

〔Use〕

Forage use; medicinal use, treating fever and cough.

① 枝条和果实 ② 枝条和花

锦鸡儿属 *Caragana* Fabr.

小叶锦鸡儿 *Caragana microphylla* Lam.

〔特征〕

多分枝灌木，高50 ~ 100cm，外皮灰黄色。长枝上的托叶宿存并硬化成针刺，长3 ~ 10mm；叶轴长15 ~ 55mm，幼时被伏柔毛，后无毛，脱落，小叶5 ~ 10对，倒卵形或矩圆形，先端obtuse，具细针尖，幼时被丝质短柔毛，长3 ~ 10mm，宽1 ~ 8mm。花梗通常单生，长10 ~ 20mm，密被丝状短柔毛，近中部有关节，每梗1花；花萼筒状钟形，被短柔毛，长9 ~ 12mm，宽5 ~ 7mm，萼齿宽三角形，长约3mm；花冠长约25mm，黄色，旗瓣圆卵形或菱形，长23 ~ 25mm；翼瓣有爪，耳短，圆齿状；龙骨瓣耳不显明，子房疏生短毛。荚果圆筒形，长4 ~ 5cm，宽5 ~ 7mm，无毛。5—6月开花，8—9月结果。

〔生境〕

生于沙丘、山地草原、干燥草原及沟谷边。

〔用途〕

饲用；还可固定流沙；药用，具有清热、收敛功效。

〔Features〕

Multi-branch shrub, 50 ~ 100cm tall, cortex grayish yellow. Stipules on the long shoots persistent and hardened into spinous, about 3 ~ 10mm long; rachises about 15 ~ 55mm long, pubescent when young, and then glabrous, deciduous, lobules 5 ~ 10 pairs, obovate or oblong, obtuse at the apex, with fine tip, silk pubescent when young, 3 ~ 10mm long, 1 ~ 8mm wide. Peduncles usually solitary, 10 ~ 20mm long, densely filiform pubescent, articulated near the middle, each peduncle 1-flowered; calyx tubular campaniform, short pubescent, 9 ~ 12mm long, 5 ~ 7mm wide ,calyx denticulate broadly triangle, about 3mm long; corolla about 25mm long, yellow, vexil ovate or rhombic, 23 ~ 25mm long; alae unguiculate, ears short, crenate; carinas inconspicuous, ovary sparsely pubescent. Legumes cylindrical, 4 ~ 5cm long, 5 ~ 7mm wide, glabrous. Flowering from May to June, and fruiting August to September.

〔Habitat〕

Sand dunes, upland meadows, dry grasslands and cheuches.

〔Use〕

Forage use; tackling quicksand; medicinal use, clearing heat, and astringent.

① 植株
② 花序
③ 叶、茎
④ 果实、叶

黄芪属 *Astragalus* Linn.

黄芪 *Astragalus membranaceus* (Fisch.) Bunge.

〔特征〕

多年生草本，高50～100cm。主根肥厚，木质，常分枝，灰白色。茎直立，上部多分枝，有细棱，被白色柔毛。羽状复叶有13～27片小叶，长5～10cm；叶柄长0.5～1cm；托叶离生，卵形，披针形或线状披针形，长4～10mm，下面被白色柔毛或近无毛；小叶椭圆形或长圆状卵形，长7～30mm，宽3～12mm，先端钝圆或微凹，具小尖头或不明显，基部圆形，上面绿色，近无毛，下面被伏贴白色柔毛。总状花序稍密，有10～20朵花；总花梗与叶近等长或较长，至果期显著伸长；苞片线状披针形，长2～5mm，背面被白色柔毛；花梗长3～4mm，连同花序轴稍密被棕色或黑色柔毛；小苞片2；花萼钟状，长5～7mm，外面被白色或黑色柔毛，有时萼筒近于无毛，仅萼齿有毛，萼齿短，三角形至钻形，长仅为萼筒的1/4～1/5；花冠黄色或淡黄色，旗瓣倒卵形，长12～20mm，顶端微凹，基部具短瓣柄，翼瓣较旗瓣稍短，瓣片长圆形，基部具短耳，瓣柄较瓣片长约1.5倍，龙骨瓣与翼瓣近等长，瓣片半卵形，瓣柄较瓣片稍长；子房有柄，被细柔毛。荚果薄膜质，稍膨胀，半椭圆形，长20～30mm，宽8～12mm，顶端具刺尖，两面被白色或黑色细短柔毛，果颈超出萼外；种子3～8颗。花期6—8月，果期7—9月。

〔生境〕

生于林缘、灌丛或疏林下，亦见于山坡草地或草甸中。

〔用途〕

药用。

〔Features〕

Perennial herb, 50 ~ 100cm tall. Taproots succulent, woody, usually branched, grey white. Stems erect, multi-branches at the top, finely carinal, white pubescent. Pinnately compound leaves have 13 ~ 27 leaflets, 5 ~ 10cm long; petioles 0.5 ~ 1cm long; stipules dialypetalous, ovoid, lanceolate or linear-lanceolate, 4 ~ 10mm long, white pubescent or glabrescent beneath; leaflets ellipse or oblong-elliptic ovoid, 7 ~ 30mm long, 3 ~ 12mm wide, obtuse or retuse at the apex, sharply cuspidate or inconspicuous, rounded at the base, green at the top, glabrescent, white pubescent beneath. Racemes slightly dense, 10 ~ 20 flowers; peduncles nearly as long as leaves or longer, significantly elongate in the fruiting period; bracts linear-lanceolate, 2 ~ 5mm long, white pubescent on the back; pedicels 3 ~ 4mm long, slightly dense brown or black pubescent with rachises; bractlets 2; calyx campaniform, 5 ~ 7mm long, white or black pubescent on the outside, occasionally calyx tube glabrescent, only calyx teeth pilose, calyx teeth short, triangular to subulate, only 1/4 ~ 1/5 as long as calyx tube; corolla yellow or faint yellow, vexils obovate, 12 ~ 20mm long, retuse at the apex, with short carpophores at the base, alae slightly shorter than vexils, limbs long elliptic, with short auricle at the base, carpophores about 1.5 times as long as limbs, carinas subequal to the alae, limbs semi-ovoid, carpophores slightly longer than limbs; ovary petiolate, finely pubescent. Legumes thin membranous, slightly inflated, semi-elliptic, 20 ~ 30mm long, 8 ~ 12mm wide, aculeate at the apex, white or black short pubescent on both sides, fruit neck beyond calyx; seeds 3 ~ 8. Flowering June to August, and fruiting July to September.

〔Habitat〕

Forest edges, bushwood or forests, hillside meadows or meadows.

〔Use〕

Medicinal use.

① 植株 ② 植株

糙叶黄芪 *Astragalus scaberrimus* Bge.

〔特征〕

多年生草本，密被白色伏贴毛。根状茎短缩，多分枝，木质化；地上茎不明显或极短，有时伸长而匍匐。羽状复叶有7～15片小叶，长5～17cm；叶柄与叶轴等长或稍长；托叶下部与叶柄贴生，长4～7mm，上部呈三角形至披针形；小叶椭圆形或近圆形，有时披针形，长7～20mm，宽3～8mm，先端锐尖、渐尖，有时稍钝，基部宽楔形或近圆形，两面密被伏贴毛。总状花序生3～5花，排列紧密或稍稀疏；总花梗极短或长达数厘米，腋生；花梗极短；苞片披针形，较花梗长；花萼管状，长7～9mm，被细伏贴毛，萼齿线状披针形，与萼筒等长或稍短；花冠淡黄色或白色，旗瓣倒卵状椭圆形，先端微凹，中部稍缢缩，下部稍狭成不明显的瓣柄，翼瓣较旗瓣短，瓣片长圆形，先端微凹，较瓣柄长，龙骨瓣较翼瓣短，瓣片半长圆形，与瓣柄等长或稍短；子房有短毛。荚果披针状长圆形，微弯，长8～13mm，宽2～4mm，具短喙，背缝线凹入，革质，密被白色伏贴毛，假2室。花期4—8月，果期5—9月。

〔生境〕

生于山坡石砾质草地、草原、沙丘及沿河流两岸的砂地。

〔用途〕

饲用；可作保持水土植物。药用，抗癌。

〔Features〕

Perennial herb, densely white adpressed hairs. Rhizomes short, multi-branched, lignified; terrestrial stems inconspicuous or very short, sometimes elongate and decumbent. Pinnately compound leaves, 7 ~ 15 lobules, 5 ~ 17cm long; petioles as long as or slightly longer than rachises; stipules lower part petioles adnate to, 4 ~ 7mm long, triangle to lanceolate at the top; lobules oval or suborbicular, occasionally lanceolate, 7 ~ 20mm long, 3 ~ 8mm wide, acute, acuminate, and sometimes slightly obtuse at the apex, broadly cuneate or suborbicular at the base, densely adpressed hairs on both surfaces. Racemes 3 ~ 5-flowered, closely arranged or slightly sparse; peduncles very short or a few centimeters lon, axillary; pedicels very short; bracts lanceolate, longer than peduncles; calyx tubular, 7 ~ 9mm long, finely adpressed hairs, calyx teeth linear lanceolate, as long as or slightly shorter than calyx tube; corolla light yellow or white, vexils obovate-oval, retuse at the apex, slightly constricted in the middle, lower part slightly narrow into inconspicuous carpophores, alae shorter than vexils, limb long elliptic, retuse at the apex, longer than carpophores, carinas shorter than alae, limb long semi-elliptic, as long as or slightly shorter than carpophores; ovary short-pilose. Legumes lanceolate long elliptic, slightly curved, 8 ~ 13mm long, 2 ~ 4mm wide, with short rostrum, dorsal suture incurvate, leathery, densely white adpressed hairs, 2 false locules. Flowering April to August and fruiting May to September.

〔Habitat〕

Hillsides, gravel grasslands, grasslands, sand dunes and sand near both sides of rivers.

〔Use〕

Forage use; can be used as plant for soil and water conservation. Medicinal use, anticancer.

① 植株 ② 叶与茎秆 ③ 花序

草木樨状黄芪 *Astragalus melilotoides* Pall.

〔特征〕

多年生草本，高60～150cm。根深长，较粗壮。茎多数由基部丛生，直立或稍斜升，多分枝，有条棱，疏生短柔毛或近无毛。单数羽状复叶，具小叶3～7；托叶三角形至披针形，基部彼此连合；叶柄有短柔毛；小叶有短柄，矩圆形或条状矩圆形，长5～15mm，宽1.5～3mm，先端钝，截形或微凹，基部截形，全缘，两面疏生白色短柔毛。总状花序腋生，比叶显著长；花小，粉红色或白色，多数。疏生，苞片甚小，锥形，比花梗短；花萼钟状，疏生短柔毛，萼齿三角形；旗瓣近圆形或宽椭圆形，基部具短爪，顶端微凹，翼瓣比旗瓣稍短，顶端2裂，基部具耳和爪，龙骨瓣比翼瓣短；子房无毛，无柄。荚果近圆形，顶端微凹，具短喙，表面有横纹，背部具稍深的沟，2室。花期7—8月。

〔生境〕

生于典型草原、森林草原以及路旁、山沟等。

〔用途〕

饲用；药用。

〔Features〕

Perennial herb, 60～150cm tall. Roots deep and long, stout. Stems numerous and fasciculate at the base, erect or slightly oblique ascending, multi-branched, carinal, sparsely short pubescent or glabrescent. Odd pinnately compound leaves, lobules 3～7; stipules triangle to lanceolate, connective at the base; petioles short pubescent; lobules have short petioles, oblong or striped-oblong, 5～15mm long, 1.5～3mm wide, obtuse, truncate or retuse at the apex, truncate at the base, the margin entire, sparsely white pubescent on both surfaces. Racemes axillary, significantly longer than leaves; flowers small, pink or white, numerous. Sparse, bracts very small, subulate, shorter than peduncles; calyx campaniform, sparsely short pubescent, calyx teeth triangle; vexil suborbicular or broadly oval, short claws at the base, retuse at the apex, alae slightly shorter than vexil, apex bifid, auricles and claws at the base, carinas shorter than alae; ovary glabrous, impedicellate. Legumes suborbicular, retuse at the apex, with short rostrum, striate on the surface, slightly sulcate on the dorsal surface, 2 locules. Flowering July to August.

〔Habitat〕

Typical grasslands, forest steppes, roadsides and valleys, etc.

〔Use〕

Forage use; medicinal use.

① 植株 ② 花 ③ 叶

斜茎黄芪 *Astragalus adsurgens* Pall.

〔特征〕

多年生草本，高20～80cm。茎多分枝并有白色或黑色毛，单数羽状复叶。具小叶7～23，小叶卵状椭圆形或椭圆，长1～3cm，宽0.5～0.8cm，背面有白毛；叶轴和小叶柄都有白色毛；托叶三角形。总状花序腋生，开多数蓝色或紫蓝色小花；花萼钟状，顶端5裂外被黑色毛；旗瓣倒卵状匙形，长约1.5cm，翼瓣长约1.2cm，龙骨瓣比翼瓣短，子房具短柄，被白毛。荚果矩圆形，长约1.5cm，顶端具短喙，表面被黑色丁字毛。花果期7—8月。

〔生境〕

生于森林草原、河滩草原、草原带、山坡草地、沟边、灌丛及林缘。

〔用途〕

饲用；绿肥；药用。

〔Features〕

Perennial herb, 20～80cm. Stems multi-branched and white or black hairs, odd pinnately compound leaves. Lobules 7～23, lobules ovate-elliptic or elliptical, 1～3cm long, 0.5～0.8cm wide, white hairs on the back; rachises and petiolules white pilose; stipules triangle. Racemes axillary, numerous blue or violet florets; calyx campaniform, apex 5-lobed covered with black hairs on the outside; vexil obovate-spathulate, about 1.5cm long, alae about 1.2cm long, carinas shorter than alae, ovary short petioles, white pilose. Legumes oblong, about 1.5cm long, short rostrum at the apex, black T-shaped hairs on the sruface. Flowering and fruiting July to August.

〔Habitat〕

Forest steppes, beach grasslands, steppe zones, hillside grasslands, ditches, bushwoods and forest edges.

〔Use〕

Forage use; green manure; medicinal use.

① 植株 ② 植株

乳白花黄芪 *Astragalus galactites* Pall.

〔特征〕

多年生草本，高5～15cm。根粗壮。茎极短缩。羽状复叶有9～37片小叶；叶柄较叶轴短；托叶膜质，密被长柔毛，下部与叶柄贴生，上部卵状三角形；小叶长圆形或狭长圆形，稀为披针形或近椭圆形，长8～18mm，宽1.5～6mm，先端稍尖或钝，基部圆形或楔形，上面无毛，下面被白色伏贴毛。花生于基部叶腋，通常2花簇生；苞片披针形或线状披针形，长5～9mm，被白色长毛；花萼管状钟形。长8～10mm，萼齿线状披针形或近丝状，长与萼筒等长或稍短，密被白色长绵毛；花冠乳白色或稍带黄色，旗瓣狭长圆形，长20～28mm，先端微凹，中部稍缢缩，下部渐狭成瓣柄，翼瓣较旗瓣稍短，瓣片先端有时2浅裂，瓣柄长为瓣片的2倍，龙骨瓣长17～20mm，瓣片短，长约为瓣柄的一半；子房无柄，有毛，花柱细长。荚果小，卵形或倒卵形，先端有喙，1室，长4～5mm，通常不外露，后期宿萼脱落，幼果有时密被白毛，后渐脱落。种子通常2颗。花期5—6月，果期6—8月。

〔生境〕

生于草原砂质土上及向阳山坡。

〔用途〕

观赏。

〔Features〕

Perennial herb, 5～15cm tall. Roots stout. Stems very short. Pinnately compound leaves have 9～37 leaflets; petioles shorter than rachis; stipules membranous, densely long pubescent, the lower part adnate to petioles, upper part oval-triangle; leaflets long elliptic or narrowly long elliptic, rarely lanceolate or nearly elliptic, 8～18mm long, 1.5～6mm wide, slightly acute or obtuse at the apex, rounded or cuneate at the base, glabrous at the top, white adpressed hairs beneath. Flowers grow in axils of the base, usually 2 flowers fascicular; bracts lanceolate or linear-lanceolate, 5 ~ 9mm long, white long-pilose; calyx tubular campaniform, 8 ~ 10mm long, calyx teeth linear-lanceolate or nearly filiform, as long as calyx tube or slightly shorter, densely white and long lanate; corollas milk white or slightly yellowish, vexil narrowly long elliptic, 20 ~ 28mm long, concave at the apex, slightly constricted in the middle, lower parts gradually narrow into carpophores, alae shorter than vexils, limbs occasionally 2-lobed at the apex, carpophores 2 times as long as limb, carinas 17 ~ 20mm long, limbs short, about half as long as carpophores; ovary impedicellate, pilose, stylus slender. Legumes small, ovoid or obvate, rostrum at the apex, 1 locule, 4 ~ 5mm long, usually not exposed, calyx deciduous in later period, fruitlets occasionally covered with dense white hairs, gradually deciduous later. Seeds usually 2. Flowering May to June and fruiting June to August.

〔Habitat〕

Grasslands, sandy soil and sunny hillside.

〔Use〕

Ornamental value.

① 植株 ② 叶 ③ 果实

甘草属 *Glycyrrhiza* Linn.

甘　草 *Glycyrrhiza uralensis* Fisch.

〔特征〕

多年生草本，高30～70cm。根和根状茎粗壮，根皮淡红色。茎直立，密被白色短毛及刺状腺体。叶为单数羽状复叶，长8～20cm，具小叶7～17，卵形或宽卵形，长2～5cm，宽1～3cm，顶端渐尖或近于钝，基部为圆形，两面被腺体及白色毛；小叶柄短，长1～1.5mm，密生白色绒毛；托叶宽披针形，被白色纤毛。总状花序腋生，长8～14mm。花密被纤毛，长14～24mm；花萼筒状，长8～14mm，有5个披针形萼齿，密被白色短毛及腺点；萼齿比萼筒稍长，上面2齿多少合生；花冠紫红色或蓝紫色，无毛，长1.4～2.5cm。荚果矩形，弯曲成镰刀状或弯曲成环状，有褐色刺状腺体，刺长1.5～2mm。种子6～8粒。6—7月开花，7—9月间结果。

〔生境〕

生于干燥草原、向阳坡、河边、田边。

〔用途〕

药用，有清肺止咳功效。

〔Features〕

Perennial herb, about 30～70cm tall. Roots and rhizomes stout, root bark faint red. Stems erect, densely white short-pilose and spinous glands. Leaves odd pinnately compound leaves, 8～20cm long, lobules 7～17, ovate or broadly ovate, 2～5cm long, 1～3cm wide, acuminate or nearly obtuse at the apex, rounded at the base, glands and white hairs on both sides; petiolules short, about 1～1.5mm long, dense white tomentum; stipules broadly lanceolate, covered with white cilium. Racemes axillary, 8～14mm long. Flowers densely covered with cilium, 14～24mm long; calyx tubular, 8～14mm long, 5 lanceolate calyx teeth, densely white short-pilose and glandular dots; calyx teeth slightly longer than calyx tube, 2 teeth somewhat concrescent; corolla aubergine or blue purple, glabrous, 1.4～2.5cm long. Legumes rectangular, bent into falciform or annular, brown spinous glands, about 1.5～2mm long. Seeds 6～8. Flowering June to July, and fruiting July to September.

〔Habitat〕

Dry grasslands, sunny lands, river banks and fields.

〔Use〕

Medicinal use, clearing lung to stop coughing.

① 植株 ② 花序 ③ 荚果 ④ 须子、叶

野豌豆属 *Vicia* Linn.

山野豌豆 *Vicia amoena* Fisch. ex DC.

〔特征〕

多年生草本，高30～100cm，植株被疏柔毛，稀近无毛。主根粗壮，须根发达。茎具棱，多分枝，细致，斜升或攀援。偶数羽状复叶，长5～12cm，几无柄，顶端卷须有2～3分枝；托叶半箭头形，长0.8～2cm，边缘有3～4裂齿；小叶4～7对，互生或近对生，椭圆形至卵披针形，长1.3～4cm，宽0.5～1.8cm；先端圆，微凹，基部近圆形，上面被贴伏长柔毛，下面粉白色；沿中脉毛被较密，侧脉扇状展开直达叶缘。总状花序通常长于叶；花10～20（30）密集着生于花序轴上部；花冠红紫色、蓝紫色或蓝色，花期颜色多变；花萼斜钟状，萼齿近三角形，上萼齿长0.3～0.4cm，明显短于下萼齿；旗瓣倒卵圆形，长1～1.6cm，宽0.5～0.6cm，先端微凹，瓣柄较宽。翼瓣与旗瓣近等长，瓣片斜倒卵形，瓣柄长0.4～0.5cm，龙骨瓣短于翼瓣，长1.1～1.2cm；子房无毛，胚珠6，花柱上部四周被毛，子房柄长约0.4cm。荚果长圆形，长1.8～2.8cm，宽0.4～0.6cm。两端渐尖，无毛。种子1～6，圆形，直径0.35～0.4cm；种皮革质，深褐色，具花斑；种脐内凹，黄褐色，长相当于种子周长的1/3。花期4—6月，果期7—10月。

〔生境〕

生于草甸、山坡、灌丛或杂木林中。

〔用途〕

饲用；药用，有去湿，清热解毒之效。

〔Features〕

Perennial herb, 30 ~ 100cm tall, sparse pubescent, glabrescent. Taproots stout, fibrous roots developed. Stems carinal, multi-branched, subtle, oblique ascending or climbing. Even pinnately compound leaves, 5 ~ 12cm long, nearly impedicellate, tendrils 2 ~ 3-branched at the apex; stipules semisagittate, 0.8 ~ 2cm long, 3 ~ 4 denticulate at the margin; lobules 4 ~ 7 pairs, alternate or nearly opposite, oval to lanceolate, 1.3 ~ 4cm long, 0.5 ~ 1.8cm wide; rounded and retuse at the apex, suborbicular at the base, long pubescent on the surface, pink-white beneath; dense hairs along the midribs, lateral veins fan-shaped unfolding to leaf margins. Racemes usually longer than leaves; 10 ~ 20(30) flowers densely adnated to the top of rachises; corolla reddish violet, blue purple or blue, color variable during flowering phase; calyx oblique campaniform, calyx teeth nearly triangle, upper calyx teeth 0.3 ~ 0.4cm long, conspicuously shorter than lower calyx teeth; vexil obovate, 1 ~ 1.6cm long, 0.5 ~ 0.6cm wide, retuse at the apex, carpophores wider. Alae approximately equal in length to vexil, limbs oblique-obovate, carpophores 0.4 ~ 0.5cm long, carinas shorter than alae, 1.1 ~ 1.2cm long; ovary glabrous, ovules 6, stylus pilose at the top, podogynium about 0.4cm long. Legumes long elliptic, 1.8 ~ 2.8cm long, 0.4 ~ 0.6cm wide. Acuminate at both ends, glabrous. Seeds 1 ~ 6, rounded, 0.35 ~ 0.4cm in diameter; testa leathery, dark brown, piebald; hilum concave, yellowish-brown, the length equivalent to one thirds of perimeter of seeds. Flowering April to June, and fruiting July to October.

〔Habitat〕

Meadows, hillsides, bushwoods or mixed forests.

〔Use〕

Forage use; medicinal use, eliminating dampness, clearing heat and removing toxicity.

① 植株 ② 花 ③ 叶、茎、花序

草木樨属 *Melilotus* Miller

白花草木樨 *Melilotus alba* Medic. ex Desr.

〔特征〕

一年生或二年生草本，高70 ~ 200cm。茎直立，圆柱形，中空，多分枝，几无毛。羽状三出复叶；托叶尖刺状锥形，长6 ~ 10mm，全缘；叶柄比小叶短，纤细；小叶长圆形或倒披针状长圆形，长15 ~ 30cm，宽（4）6 ~ 12mm，先端钝圆，基部楔形，边缘疏生浅锯齿，上面无毛，下面被细柔毛，侧脉12 ~ 15对，平行直达叶缘齿尖，两面均不隆起，顶生小叶稍大，具较长小叶柄，侧小叶小叶柄短。总状花序长8 ~ 20cm，腋生，具花40 ~ 100朵，排列疏松；苞片线形，长1.5 ~ 2mm；花长4 ~ 5mm；花梗短，长1 ~ 1.5mm；萼钟形，长约2.5mm，微被柔毛，萼齿三角状披针形，短于萼筒；花冠白色，旗瓣椭圆形，稍长于翼瓣，龙骨瓣与翼瓣等长或稍短；子房卵状披针形，上部渐窄至花柱，无毛，胚珠3 ~ 4粒。荚果椭圆形至长圆形，长3 ~ 3.5mm，先端锐尖，具尖喙，表面脉纹细，网状，棕褐色，老熟后变黑褐色；有种子1 ~ 2粒。种子卵形，棕色，表面具细瘤点。花期5—7月，果期7—9月。

〔生境〕

生于田边、路旁荒地及湿润的砂地。

〔用途〕

饲用；药用，具有清热解毒、化湿杀虫功效。

〔Features〕

Annual and biennial herb, 70 ~ 200cm tall. Stems erect, cylindrical, hollow, multi-branched, glabrescent. Pinnate ternately compound leaves; stipules spiny subulate, 6 ~ 10mm long, the margin entire; petioles shorter than lobules, slender; lobules long elliptic or lanceolate-ovale, 15 ~ 30cm long, (4) 6 ~ 12mm wide, obtuse at the apex, cuneate at the base, sparsely serrulate at the margin, glabrous at the top, finely pubescent, lateral veins 12 ~ 15 pairs, parallel to cusp of leaf margins, not ridgy on both surfaces, terminal lobules slightly larger, longer petiolules, petiolules shorter. Racemes 8 ~ 20cm long, axillary, 40 ~ 100 flowers, loosely arranged; bracts linear, 1.5 ~ 2mm long; flowers 4 ~ 5mm long; peduncles short, about 1 ~ 1.5mm long; calyx campaniform, about 2.5mm long, puberulent, calyx teeth triangular lanceolate, shorter than calyx tube; corolla white, vexil oval, slightly longer than alae, carinas as long as or slightly shorter than alae; ovary ovate-lanceolate, gradually narrow to stylus, glabrous, ovules 3 ~ 4. Legumes oval to long elliptic, 3 ~ 3.5mm long, acuminate at the apex, acute rostrum, finely veining on the surface, reticulate, brown, black brown when mature; seeds 1 ~ 2. Seeds ovate, brown, finely tuberculate on the surface. Flowering May to July, and fruiting July to September.

〔Habitat〕

Fields, roadsides, wastelands and wet sandy ground.

〔Use〕

Forage use; medicinal use, clearing heat and removing toxicity, eliminating dampness and pesticidal.

① 植株 ② 叶 ③ 花

扁蓿豆属 *Medicago* Linn.

花苜蓿 *Medicago ruthenica*（Linn.）Trautv.

〔特征〕

多年生草本，高20 ~ 100cm。茎、枝四棱形，疏被白色柔毛。叶为羽状三出复叶，叶柄长5mm，互生；中间小叶卵形。倒卵形或倒披针形，长5 ~ 12mm，宽3 ~ 7mm，顶端圆形或截形，并微凹或具刺，上部边缘有锯齿；两边小叶较小；叶柄和小叶背面被纤细白毛；托叶披针形，具脉状纹。总状花序腋生，花序梗长约1mm，小花3 ~ 10(12)朵；花萼钟状，长约3mm，被白毛，具5萼齿，三角形；花冠黄色，有紫色脉；子房条形。荚果扁平条状矩圆或卵形，长7 ~ 10mm，宽5mm，先端短尖，有网纹，种子2 ~ 4粒。矩圆状椭圆形。花果期7—9月。

〔生境〕

生于草原、向阳坡、路旁、河滩沙质地。

〔用途〕

饲用；药用，具有清热解毒、止咳、止血功效。

〔Features〕

Perennial herb, 20 ~ 100cm tall. Stems and branches quadrangual prism, sparsely white pubescent. Pinnate ternately compound leaves, petioles 5mm long, alternate; the middle lobules ovate. Obovate or oblanceolate, 5 ~ 12mm long, 3 ~ 7mm wide, rounded or truncate at the apex, retuse or aculeate, the upper margins serrulate; lobules smaller on both sides; the back of petioles and lobules finely tomentose; stipules lanceolate, veniform. Racemes axillary, peduncles about 1mm long, florets 3 ~ 10 (12); calyx campaniform, about 3mm long, white pilose, 5 calyx teeth, triangle; corolla yellow, purple veins; ovary striped. Legumes flat striped rounded-rectangular or ovate, 7 ~ 10mm long, 5mm wide, mucroniform at the apex, reticulate patterns, seeds 2 ~ 4. rounded-rectangular oval. Flowering and fruiting July to September.

〔Habitat〕

Grasslands, sunny lands, roadsides, beachlands and sandy lands.

〔Use〕

Forage use; medicinal use, clearing heat and removing toxicity, cough relieving, and arresting bleeding.

① 植株
② 叶
③ 花

黄华属　*Thermopsis* R. Br.

披针叶黄华　*Thermopsis lanceolata* R. Br.

〔特征〕

多年生草本，高20～45cm，密被白色伏毛。三出复叶，互生；小叶矩圆状倒卵形或倒披针形，长2.5～8.5cm，宽1～20mm，顶端尖，基部楔形，全缘，背面密被短柔毛；托叶两，卵状披针形，基部彼此连合。总状花序顶生；花序轴每节轮生2～3朵黄色花；苞片轮生，密被伏柔毛，基部彼此连合；花萼钟状，长1.6～1.8cm，被短柔毛，裂片5枚；花瓣蝶状，旗瓣圆形。荚果条形，长5～7cm，宽7～12mm，具细长喙，密被短毛；种子6～14粒，肾状，光亮黑褐色。花期5—7月，果期7—10月。

〔生境〕

生于河边、路旁、沙质地、轻碱性沙质土壤上。

〔用途〕

药用，能祛痰、镇咳；可杀蛆。

〔Features〕

Perennial herb, 20～45cm tall, densely white adpressed hairs. Ternately compound leaves, alternate; lobules rounded-rectangular obovate or oblanceolate, 2.5～8.5cm long, 1～20mm wide, acute at the apex, cuneate at the base, the margin entire, densely short pubescent on the back; stipules 2, ovate-lanceolate, connective at the base. Racemes terminal; each internode of rachis verticillate 2～3 yellow flowers; bracts verticillate, densely pubescent, connective at the base; calyx campaniform, 1.6～1.8cm long, short pubescent, lobes 5; petals papilionaceus, vexil rounded. Legumes striped, 5～7cm long, 7～12mm wide, long and thin rostrum, densely short-pilose; seeds 6～14, reniform, lustrous, black brown. Flowering May to July and fruiting July to October.

〔Habitat〕

River banks, roadsides, sandy lands, light alkaline sandy soils.

〔Use〕

Medicinal use, eliminating phlegm, cough relieving; and killing maggots.

① 植株1 ② 植株2 ③ 茎、叶 ④ 花

棘豆属 *Oxytropis* DC.

多叶棘豆 *Oxytropis myriophylla*（Pall）DC

〔特征〕

多年生草本，高20～30cm，全株被白色或黄色长柔毛。根褐色，粗壮，深长。茎缩短，丛生。轮生羽状复叶长10～30cm；托叶膜质，卵状披针形，基部与叶柄贴生，先端分离，密被黄色长柔毛；叶柄与叶轴密被长柔毛；小叶25～32轮，每轮4～8片或有时对生，线形、长圆形或披针形，长3～15mm，宽1～3mm，先端渐尖，基部圆形，两面密被长柔毛。多花组成紧密或较疏松的总状花序；总花梗与叶近等长或长于叶，疏被长柔毛；苞片披针形，长8～15mm，被长柔毛；花长20～25mm；花梗极短或近无梗；花萼筒状，长11mm，被长柔毛，萼齿披针形，长约4mm，两面被长柔毛；花冠淡红紫色，旗瓣长椭圆形，长18.5mm，宽6.5mm，先端圆形或微凹，基部下延成瓣柄，翼瓣长15mm，先端急尖，耳长2mm，瓣柄长8mm，龙骨瓣长12mm，喙长2mm，耳长约15.2mm；子房线形，被毛，花柱无毛，无柄。荚果披针状椭圆形，膨胀，长约15mm，宽约5mm，先端喙长5～7mm，密被长柔毛，隔膜稍宽，不完全2室。花期5—6月，果期7—8月。

〔生境〕

生于砂地、平坦草原、干河沟、丘陵地、轻度盐渍化沙地及石质山坡或低山坡。

〔用途〕

饲用；药用，有清热解毒、消肿、祛风湿、止血之功效。

〔Features〕

Perennial herb, 20 ~ 30cm tall, the whole plant is covered with white or yellow pubescent. Roots brown, stout, long. Stems shortened, fasciculate. Verticillate pinnately compound leaves 10~30cm long; stipules membranous, ovate-lanceolate, adnate to petioles at the base, separated at the apex, densely yellow pubescent; petioles and rachises are densely covered with long pubescent; leaflets 25 ~ 32 whorls, each whorl 4 ~ 8 pieces or occasionally opposite, linear, long elliptic or lanceolate, 3 ~ 15mm long, 1 ~ 3mm wide, acuminate at the apex, rounded at the base, densely long pubescent on both sides. Many flowers are arranged into dense or sparse racemes; peduncles nearly as long as leaves or longer than leaves, sparsely long pubescent; bracts lanceolate, 8 ~ 15mm long, long pubescent; flowers 20 ~ 25mm long; pedicels very short or nearly sessile; calyx tubular, 11mm long, long pubescent, calyx teeth lanceolate, about 4mm long, long pubescent on both sides; corolla reddish purple, vexils long elliptic, 18.5mm long, 6.5mm wide, rounded or retuse at the apex, decurrent into carpophores at the base, alae 15mm long, acute at the apex, auricle 2mm long, carpophores 8mm long, carinas 12mm long, rostrum 2mm long, ears about 15.2mm long; ovary linear, pilose, stylus glabrous, impedicellate. Legumes lanceate-elliptic, inflated, about 15mm long, about 5mm wide, rostrum 5 ~ 7mm long at the apex, densely pubescent, diaphragm slightly wide, incomplete 2 locules. Flowering May to June, and fruiting July to August.

〔Habitat〕

Sandy ground, flat grasslands, dry streams, hills, mild saliferous sand and stony hillside or hillside.

〔Use〕

Forage use; medicinal use, clearing heat and antidotic, mitigating edema, expelling wind-damp, and hemostatic.

① 植株 ② 花蕾 ③ 叶 ④ 花

牻牛儿苗科 Geraniaceae

牻牛儿苗属 *Erodium* L′ H é r.

牻牛儿苗 *Erodium stephanianum* Willd.

〔特征〕

一年生或二年生草本，高30～50cm，根细长。茎细弱，平卧或稍斜升，多分枝，被白色毛，有膨大的关节。叶对生，长卵形或椭圆状三角形，长约6cm，二回羽状深裂；最末小叶条状，顶端渐尖，具1～3长锯齿或全缘，两面密被纤柔毛；基生叶叶柄被白毛，长约10cm。夏季开花，花小，蓝紫色，成腋生伞形花序，小花2～6朵，花序梗长5～15cm；苞片披针形，边缘有毛；萼片5裂，背面密生白毛。顶端具细尖；花冠5枚，蓝紫色；雄蕊10，花柱基部膨大，外圈5枚无花丝；蜜腺点5个，黄色；蒴果具长喙，长3～4cm，无毛，熟时基部开裂，分为5个分果，有螺旋状嘴。

〔生境〕

生于草原、山坡、田边、路旁、宅旁。

〔用途〕

药用，具有强筋骨、祛风湿、清热解毒功效。

〔Features〕

Annual or biennial herb, 30～50cm tall, roots long and thin. Stems thin, prostrate or slightly oblique ascending, multi-branched, white pilose, inflated articulated. Leaves opposite, long ovate or elliptical triangle, about 6cm long, bipinnatiparted; terminal lobules striped, acuminate at the apex, 1～3-long serrate or the margin entire, densely pubescent on both sides; petioles of basal leaves white pilose, about 10cm long. Blooming in summer, flowers small, blue purple, axillary umbels, florets 2～6, peduncles 5～15cm long; bracts lanceolate, pilose at the margin; sepals 5-lobed, densely white pilose on the back. Apiculus at the apex; corollas 5, blue purple; stamens 10,stylus inflated at the base, outer 5 ones no filament; nectary glands 5, yellow; capsule long rostrum, 3～4cm long, glabrous, cracked at the base when mature, 5 schizocarps with heliciform mouth.

〔Habitat〕

Grasslands, hillsides, fields, roadsides, next to residential houses.

〔Use〕

Medicinal use, strengthening the bones and muscles, expelling wind-damp, clearing heat and removing toxicity.

① 植株 ② 花 ③ 果实 ④ 花与株丛

亚麻科 Linaceae

亚麻属 *Linum* L.

宿根亚麻 *Linum perenne* L.

〔特征〕

多年生草本，高20～70cm。主根垂直，粗壮，木质化。茎从基部丛生，直立或稍斜生，分枝，通常有无不育枝。叶互生，条形或条状披针形，长1～1.3mm，宽1～3cm，基部狭窄，先端尖，具1脉，平或边缘稍尖，无毛；下部叶有时较小，鳞片状；不育枝上的叶较密，条形，长7～12mm，宽0.5～1mm。聚伞花序，花通常多数，暗蓝色或蓝紫色，直径约2cm，花梗细长，稍弯曲，偏向一侧，长1～2.5cm；萼片卵形，下部有5条突出的脉，边缘膜质，先端尖；花瓣倒卵形，长约1cm，基部楔形；雄蕊与花柱异长，稀等长。蒴果近球形，直径6～7mm，草黄色，开裂；种子矩圆形，栗色。

〔生境〕

广泛生于草原、山坡地带。

〔用途〕

种子可榨油；药用，具有通经利尿功效。

〔Features〕

Perennial herb, 20～70cm tall. Taproots vertical, stout, lignified. Stems fasciculate at the base, erect or slightly oblique, branched, usually sterile branches. Leaves alternate, striped or striped-lanceolate, 1～1.3mm long, 1～3cm wide, narrow at the base, acute at the apex, 1 veined, flat or slightly acute at the margin, glabrous; bottom leaves sometimes small, scaly; leaves dense on the sterile branches, striped, 7～12mm long, 0.5～1mm wide. Cymes and flowers usually numerous, dark blue or blue purple, about 2cm in diameter, peduncles long and thin, slightly curved, inclined to one side, 1～2.5cm long; sepals ovate, 5 protruding veins at lower part, membranous margins, acute at the apex; petals obovate, about 1cm long, cuneate at the base; stamens and heterostyly rarely equilong. Capsules subsphaeroidal, 6～7mm in diameter, straw yellow, cracked; seeds oblong, maroon.

〔Habitat〕

Grasslands and hillsides.

〔Use〕

Seeds can be used for oil manufacture; medicinal use, restoring menstrual flow and diuretic.

① 植株 ② 花 ③ 株丛

芸香科 Rutaceae

拟芸香属 *Haplopyllum* A.Juss.

北芸香 *Haplophyllum dauricum* (L.) G.Don.

〔特征〕

多年生草本，高15～30cm，基部木质化。茎、枝很多，丛生；小支纤细，有腺点。叶互生，无柄，条状披针形或条状矩圆形至倒卵形之间，长0.5～1.5cm，宽1～2.5mm，顶端渐尖，基部楔形，全缘，具有腺点。伞房状聚伞花序顶生，具两性花3～20朵，小苞片条形；萼片5枚，宽卵形；花瓣5枚，黄色，矩圆形，长6～8mm，全缘；雄蕊10，花丝下部有翅状长白毛；雌蕊4～5枚，但是中部以下彼连合；子房球形，无柄，2～4室，花柱细长。果实为蒴果；每室含有2粒肾状深褐色种子。

〔生境〕

生于干燥草原、山坡草地、路旁、碱性土壤上。

〔用途〕

药用，具有通经、祛风功效。

〔Features〕

Perennial herb, 15～30cm tall, lignified at the base. Stems and branches numerous, fasciculate; branchlets slender, glandular dots. Leaves alternate, impedicellate, striped lanceolate or oblong to obovate, 0.5～1.5cm long, 1～2.5mm wide, acuminate at the apex, cuneate at the base, the margin entire, with glandular dots. Corymbose cymes terminal, bisexual flowers 3～20, bractlets striped; sepals 5, broadly ovoid; petals 5, yellow, rounded-rectangular, 6～8mm long, the margin entire; stamens 10, aliform long white polise at lower part of filaments; pistil 4～5, but connective below the middle part; ovary spherical, impedicellate, 2～4 locules, stylus long and thin. Fruit is capsule; each locule containing 2 reniform dark brown seeds.

〔Habitat〕

Dry grasslands, hillside grasslands, roadsides and alkaline soil.

〔Use〕

Medicinal use, restoring menstrual flow and dispelling wind evil.

① 植株1　② 植株2　③ 叶　④ 花、花瓣

远志科　Polygalaceae

远志属　*Polygala* Linn.

远　志　*Polygala tenuifolia* Willd.

〔特征〕

多年生草本，根长肉质。茎较细，多分枝，高25 ~ 35cm，无毛或近无毛。叶互生，披针形或条形，长1 ~ 3cm，宽1.5 ~ 3mm，顶端渐尖，基部狭，全缘。短总状花序顶生，苞片3枚；萼片5枚，宿存，内边两枚裂片较大，花瓣状；花瓣3枚，中间1枚较大，龙骨瓣状，上部有鸡冠状物；雄蕊8，花丝彼此连合成管状；子房上位；花柱弯曲，柱头2个，不等长。蒴果扁倒卵形，中央有纵沟。边缘无睫毛；种子2粒，卵形，黑色并密被白毛。

〔生境〕

生于山丘、向阳坡、沙质草地、路旁、河边、田边等地。

〔用途〕

药用，治失眠、惊悸、咳嗽多痰等症。

〔Features〕

Perennial herb, roots long and succulent. Stems slender, multi-branched, 25 ~ 35cm tall, glabrous or glabrescent. Leaves alternate, lanceolate or striped, 1 ~ 3cm long, 1.5 ~ 3mm wide, acuminate at the apex, narrow at the base, the margin entire. Short racemes terminal, bracts 3; sepals 5, persistent, inner 2 lobes larger, petaloid; petals 3, middle one larger, keel-shaped, cristate at the top; stamens 8, filament unite into tubular; ovary superior; stylus curved, stigmas 2, unequal length. Capsules flat obovate, longitudinal furrows in the center. Non-ciliiform at the margin; seeds 2, ovate, black and densely white pilose.

〔Habitat〕

Hills, sunny lands, sandy grasslands, roadsides, river banks and fields.

〔Use〕

Medicinal use, treating insomnia, palpitation and cough accompanied by excessive phlegm.

① 植株1 ② 植株2 ③ 花、果实 ④ 叶 ⑤ 叶

大戟科 Euphorbiaceae

大戟属 *Euphorbia* Linn.

乳浆大戟 *Euphorbia esula* L.

〔特征〕

多年生草本，15 ~ 40cm，含乳汁，全株稍有毛。茎直立，具纵纹，下部浅紫色，上部2 ~ 5分枝；枝上叶多，叶条形，长1.5 ~ 3cm；叶互生，倒披针形或条状披针形，顶端钝，基部楔形。顶生花序具3 ~ 5总梗；总梗再次分2 ~ 3枝；苞片对生，宽心形，全缘，顶端突出；苞片间生有杯状花序；总苞片黄色，边缘具有两头有4个月牙状的腺点；雄蕊多数，雌蕊1枚；子房具长柄，3室，花柱3个；花无花被。蒴果无毛；种子光滑，卵形，长约2mm，灰褐色或具棕色点纹。

〔生境〕

生于山坡草地、山沟、草原上。

〔用途〕

药用；亦可作农药。

〔Features〕

Perennial herb, 15 ~ 40cm, latiferous, the whole plant is slightly hairy. Stems erect, with longitudinal striations, lower part shallow purple, 2 ~ 5-branched at the top; many leaves on the branches, leaves striped, 1.5 ~ 3cm long; leaves alternate, oblanceolate or striped lanceolate, obtuse at the apex, cuneate at the base. Terminal inflorescence has 3 ~ 5 peduncles; peduncles 2 ~ 3-branched; bracts opposite, broadly heart-shaped, the margin entire, apex protruding; bracts cyathiform inflorescences; phyllary yellow, 4 crescent glandular dots at the margin; stamens numerous, pistil 1; ovary long petioles, trilocular, stylus 3; flowers achlamydeous. Capsules glabrous; seeds glabrous, ovate, about 2mm long, taupe or with brown dots.

〔Habitat〕

Hillside grasslands, valleys and grasslands.

〔Use〕

Medicinal use; also can be used as pesticide.

① 植株 ② 开花期株丛 ③ 花 ④ 开花期景观

瑞香科 Thymelaeaceae

狼毒属 *Stellera* Linn.

狼 毒 *Stellera chamaejasme* L.

〔特征〕

多年生草本，高20～50cm。根木质化粗壮圆柱形，少分枝或不分枝。茎多数，丛生，直立，无分枝，无毛，基部木质化。叶互生，无柄，椭圆状披针形或披针形，无毛，全缘，长1.4～2.8cm，宽3～9mm，头状花序顶生，总苞片绿色；花被筒纤细，长8～12mm，紫红色并有纵脉，顶端5裂，裂片长2～3mm，乳白色并有紫红色网纹；雄蕊10，花被筒内侧两轮生，花丝极短；子房一室，上部有淡黄色纤柔毛，柱头球状。果实包裹在宿存花被内。

〔生境〕

生于沙质草原、山地向阳坡、平地草原及河滩上。

〔用途〕

药用，行水、消积；亦可作杀虫药。

〔Features〕

Perennial herb, 20～50cm tall. Roots lignified, stout, cylindrical, subramous or unbranched. Stems numerous, fasciculate, erect, no branches, glabrous, lignified at the base. Leaves alternate, impedicellate, elliptical lanceolate or lanceolate, glabrous, the margin entire, 1.4～2.8cm long, 3～9mm wide, capitellate inflorescence terminal, phyllary green; perianth tube slender, 8～12mm long, aubergine and longitudinal veins, apex 5-lobed, lobes 2～3mm long, milk white and aubergine reticulate patterns; stamens 10,verticillate on the inner side of perianth tube, filaments very short; ovary unilocular, light yellow finely pubescent at the top, stigmas globose. The fruit wrapped in persistent perianth.

〔Habitat〕

Sandy grasslands, mountain lands, sunny lands, flat grasslands and beachlands.

〔Use〕

Medicinal use, dissipation of pathogenic water, removing food retention; and also can be used as pesticide.

① 植株 ② 花序 ③ 花蕾 ④ 叶

伞形科 Umbelliferae

迷果芹属 *Sphallerocarpus* Bess. ex DC.

迷果芹 *Sphallerocarpus gracilis* (Bess.) K. Pol.

〔特征〕

二年生草本，高50～120cm。茎下部疏生毛和关节密被硬毛。茎生叶卵状，长5～15cm，三角状羽状复叶；小叶披针形至条状披针形之间，长5～10mm，宽1～3mm，羽状深裂，背面沿脉疏生柔毛；叶柄长1.5～8cm，疏生硬毛。复伞形花序顶生，花序总柄长1.5～6cm，总苞片一个或无总苞片；小总苞片5，披针形，边缘具睫毛，小伞形花序5～10，每个花序含10～15朵白花，花柄长5～10mm。双悬果矩圆状椭圆形，长3～6mm，宽1～2mm，无毛，扁平，两侧压扁。

〔生境〕

生于山坡草地、路边、林边、河边及草甸草原。

〔用途〕

饲用；食用；药用，具有祛肾寒、敛黄水功效。

〔Features〕

Biennial herb, 50～120cm tall. Lower part of stems sparsely pilose and the joints densely strigose. Cauline leaves oval, 5～15cm long, triangular pinnately compound leaves; lobules lanceolate to striped lanceolate, 5～10mm long, 1～3mm wide, pinnateparted, sparsely pubescent along veins on the back; petioles 1.5～8cm long, sparsely strigose. Compound umbels terminal, peduncles 1.5～6cm long, phyllary 1 or no phyllary; small phyllaries 5, lanceolate, ciliiform at the margin, small umbels 5～10, each inflorescence containing 10～15 white flowers, pedicel 5～10mm long. Mericarps rounded-rectangular oval, 3～6mm long, 1～2mm wide, glabrous, flat, oblate on both sides.

〔Habitat〕

Hillside meadows, roadsides, forest edges, river banks and meadow steppes.

〔Use〕

Forage use; edible use; medicinal use, dispelling cold and astringent.

① 植株 ② 花序 ③ 花 ④ 叶

柴胡属 *Bupleurum* L.

狭叶柴胡 *Bupleurum scorzonerifolium* Willd.

〔特征〕

多年生草本，高30～60cm，茎上部分枝。主根细长螺旋状，黄褐色。茎单生或2～3丛生，基部有红色老叶的残留物。叶互生，基部叶和茎下部叶有长柄，条形或细条形，长6～16cm，宽2～7mm，顶端尖，基部骤尖，两面有平行5～7纵脉。复伞形花序多数，每个有3～8小伞形花组成；总苞片1～3，针形；小伞形花序开6～15朵黄色小花；小总苞片5枚，细披针形；花小，具弯曲的5枚花瓣，雄蕊5，花药卵状，花柱极短。双悬果长圆柱形或宽椭圆形，长2.5mm，宽2mm，粗壮并有5个钝棱。

〔生境〕

生于草原、丘陵坡地、固定沙丘、山坡草地及灌丛中。

〔用途〕

药用，具有解表退热、疏肝解郁功效。

〔Features〕

Perennial herb, 30～60cm tall, upper part of stems branched. Taproots elongated heliciform, yellowish-brown. Stems solitary or 2～3 fasciculate, residue of red old leaves at the base. Leaves alternate, basal leaves and leaves at lower part of stems have long petioles, striped or finely striped, 6～16cm long, 2～7mm wide, abruptly acute at the apex, cuspidate at the base, parallel 5～7 longitudinal veins on both sides. Compound umbels numerous, each consisting of 3～8 umbellate flowers; phyllaries 1～3, aciculiform; small umbels blossom 6～15 yellow florets; small phyllaries 5, finely lanceolate; flowers small, with curved 5 petals, stamens 5, anthers oval, stylus ultrashort. Mericarp long cylindrical or broadly oval, 2.5mm long, 2mm wide, stout and 5 obtuse carinal.

〔Habitat〕

Grasslands, hills sloping fields, fixed sand dunes, hillside grasslands and bushwoods.

〔Use〕

Medicinal use, relieving exterior syndrome and antipyretic and discharging liver and relieving depression.

① 植株 ② 花序、花蕾 ③ 花序 ④ 茎、叶

防风属 *Saposhnikovia* Schischk.

防　风 *Saposhnikovia divaricata* (Turcz.) Schischk.

〔特征〕

多年生草本，高30～80cm。全株无毛。根粗壮，根上部和茎下部有许多棕黄色叶柄的残留物。茎分枝呈双叉式。基生叶变成叶鞘状抱茎的2～6.5cm长叶柄，三角状卵形，长7～19cm，一至二回或三回羽状全裂；末回小叶片条形或披针形，长4～5mm，宽1～9mm，全缘或疏缺刻，复伞形花序多数，直径1.5～3.5cm，总梗长2～5cm，无总苞片或稀一片；小伞形花序5～9，每小伞形花序含4～9朵黄色或白色花；小总苞片4～5裂，条形至披针形之间。双悬果圆柱状宽卵形，上有瘤点，长3～5mm，宽2～2.5mm，扁平，侧棱具翅；种子长圆柱形。

〔生境〕

生于草原、丘陵地带、山坡草地、田边、路旁。

〔用途〕

药用。

〔Features〕

Perennial herb, 30～80cm tall. The whole plant glabrous. Roots stout, a lot of residue of brownish yellow petioles at the top of roots and lower part of stems. Stems branched and bi-fork. Basal leaves turn into sheath shape amplexicaul 2～6.5cm-long petioles, triangular ovate, 7～19cm long, unipinnate to bipinnate or tripinnate lobed; ultimate leaf-blades striped or lanceolate, 4～5mm long, 1～9mm wide, the margin entire or sparsely erose, compound umbels numerous, 1.5～3.5cm in diameter, peduncles 2～5cm long, no phyllary or sparse; small umbels 5～9, each umbel containing 4～9 yellow or white flowers; small phyllaries 4～5-lobed, striped to lanceolate. Mericarps cylindrical broadly ovate, tuberculate, 3～5mm long, 2～2.5mm wide, flat, lateral edges epipterous; seeds long cylindrical.

〔Habitat〕

Grasslands, hilly lands, hillside grasslands, fields and roadsides.

〔Use〕

Medicinal use.

① 植株　② 花序　③ 花

白花丹科　Plumbaginaceae

补血草属　*Limonium* Mill.

二色补血草　*Limonium bicolor* (Bunge) Kuntze.

〔特征〕

多年生草本，高20 ~ 70cm，根肉质，根皮红褐色。茎直立或斜升，上部多分枝开展；全株无毛。基生叶丛生，匙形或倒卵形，长2 ~ 8cm，宽1 ~ 2.5cm，全缘，顶端急尖，基部长楔形，通常棕紫色，疏生腺点；茎下部叶条状倒披针形，上部叶长三角状鳞片形。密集簇生的多数聚伞花序集成大圆锥状花序；花2 ~ 3朵集成小穗，每朵花有2枚苞片；苞片宽卵形，紫红色或淡紫绿色，边缘膜质；花萼漏斗状，长6 ~ 8mm，干膜质；喉部密生毛并裂片5，白色或粉红色，宿存；花瓣5枚，黄色，生在喉部内并外伸，基部合生，先端深裂，宿存；雄蕊5，生于花瓣基部，子房上位；花柱5个，分离，无毛。果实有5棱，包裹在宿存花被内。花果期5—8月。

〔生境〕

散生于草甸草原、沟谷边、河边、盐碱滩上。

〔用途〕

药用，具有止血散瘀功效。

〔Features〕

Perennial herb, 20 ~ 70cm tall, roots succulent, root bark russet. Stems erect or oblique ascending, multi-branched squarrose at the top; the whole plant glabrous. Basal leaves fasciculate, spathulate or obovate, 2 ~ 8cm long, 1 ~ 2.5cm wide, the margin entire, acute at the apex, long cuneate at the base, usually purplish brown, sparsely glandular dots; leaves striped oblanceolate at lower part, leaves long triangular squamelliform at the top. Densely fascicular cymes integrate into conical inflorescence; 2 ~ 3 flowers clustered into a spikelet, each flower has 2 bracts; bracts broadly ovate, aubergine or lavender green, membranous margins; calyx funnel-shaped, 6 ~ 8mm long, scarious; throat densely pilose and lobes 5, white or pink, persistent; petals 5, yellow, born in the throat and lituate, concrescent at the base, apex parted, persistent; stamens 5, born at the base of petals, ovary superior; stylus 5, separated, glabrous. Fruit 5 carinal, wrapped in the persistent perianth. Flowering and fruiting May to August.

〔Habitat〕

Meadow steppes, cheuches, river banks and salt marshes.

〔Use〕

Medicinal use, arresting bleeding and dispersing blood stasis.

① 植株 ② 花与梗 ③ 花序 ④ 花

龙胆科 Gentianaceae

龙胆属 *Gentiana*（Tourn.）

达乌里龙胆 *Gentiana dahurica* Fisch.

〔特征〕

多年生草本，高15～30cm。根黄褐色，索状。茎直立或斜升。有四棱角。上部少分枝，基部有宿存老叶，基部叶丛生，披针形，先端渐尖，基部楔形，全缘，有3条平行主脉。叶对生，卵形至披针形，基部抱茎。聚伞花序腋生，具1～8朵花，总花梗细长；花萼，花冠钟状，裂片各5枚；花深蓝色；雄蕊5，花丝基部有宽翅；花柱短，柱头两裂。蒴果矩圆形。种子条形，边缘有翅。

〔生境〕

生于山坡草丛、山沟、平原、丘陵地带。

〔用途〕

药用，具有泻肝胆实火、清湿热功效。

〔Features〕

Perennial herb, 15～30cm tall. Roots yellowish-brown, funicular. Stems erect or oblique ascending, 4 corner angles. A few branches at the top, persistent old leaves at the base, basal leaves fasciculate, lanceolate, acuminate at the apex, cuneate at the base, the margin entire, 3 parallel primary veins. Leaves opposite, ovate to lanceolate, amplexicaul at the base. Cymes axillary, 1～8 flowers, peduncles elongated; calyx and corolla campaniform, lobes 5 each; flowers dark blue; stamens 5, wide winged at the base of filaments; stylus short, stigmas bifid. Capsules oblong. Seeds striped, winged at the margin.

〔Habitat〕

Hilly bushes, valleys, plains, and hills.

〔Use〕

Medicinal use, purging liver-gallbladder fire, eliminating dampness and heat.

① 植株 ② 花

旋花科 Convolvulaceae

打碗花属 *Calystegia* R.Br.

打碗花 *Calystegia hederacea* Wall. ex Roxb.

〔特征〕

一年生缠绕或平卧草本，全体无毛。具细长白色的根茎。茎具细棱；通常有基部分枝。叶片三角状卵形、戟形或箭形，侧面裂片尖锐，近三角形，或2～3裂，中裂片矩圆形或矩圆状披针形，长2～4.5(5)cm，基部（最宽处）宽(1.7)3.5～4.8cm，先端渐尖，基部微心形，全缘，两面通常无毛。花单生叶腋，花梗长于叶柄，有细棱；苞片宽卵形，长7～11(16)mm；花冠漏斗状，淡粉红色或淡紫色，直径2～3cm；雄蕊花丝基部扩大，有细鳞毛；子房无毛，柱头2裂，裂片矩圆形，扁平。蒴果卵圆形，微尖，光滑无毛。花期7—9月；果期8—10月。

〔生境〕

生于撂荒地、路旁、溪边。

〔用途〕

药用；饲用；根茎可造酒，也可制饴糖。

〔Features〕

Annual twining or prostrate herb, all glabrous. Elongated white rhizomes. Stems finely carinal; usually branched at the base. Leaf-blades triangular-ovate, hastate or sagittal, lobes acute on side face, nearly triangle, or 2～3-lobed, midlobes oblong or rounded-rectangular lanceolate, 2～4.5(5)cm long, (1.7)3.5～4.8cm wide at the base (the widest part), acuminate at the apex, heart-shaped at the base, the margin entire, usually glabrous on both surfaces. Flowers solitary axil, peduncles longer than petioles, finely carinal; bracts broadly ovate, 7～11(16)mm long; corolla funnel-shaped, rose pink or lavender, 2～3cm in diameter; the base of staminal filaments expanded, with finely scaly hairs; ovary glabrous, stigmas bifid, lobes oblong, flat. Capsules oval, acutate and glabrous. Flowering July to September; and fruiting August to October.

〔Habitat〕

Wastelands, roadsides and by the streams.

〔Use〕

Medicinal use; forage use; rhizomes can be used to make wine and malt sugar.

① 植株 ② 小花 ③ 花瓣

紫草科 Boraginaceae

齿缘草属 *Eritrichium* Schrad.

石生齿缘草 *Eritrichium rupestre* (Pall.) Bunge

〔特征〕

多年生草本，高10～30cm，密生灰色糙伏毛，基部多分枝丛生，基生叶丛生，匙形或匙状条形，全缘；茎生叶互生，无柄，倒披针形至条状披针形，先端尖；各叶都被灰色糙伏毛；花较小，淡蓝色，无梗，花单生或几朵顶生总状花序，苞片条形；花萼条状五裂，花蓝色，花冠喉部有五个附属物；雄蕊5，柱头头状；花冠钝，五裂；花冠管状或漏斗状；花盘存在；子房上位，由两心皮组成，每一心皮有胚珠两粒。果实为4个小坚果陀螺形，有小疣状突和5～100个小齿。

〔生境〕

生于山顶、石质山坡草地、砾石缝或路边。

〔用途〕

药用，具有清热鲜毒功效。

〔Features〕

Perennial herb, 10～30cm tall, densely gray strigose, multi-branched fasciculate at the base, basal leaves fasciculate, spathulate or spatulate striped, the margin entire; cauline leaves alternate, impedicellate, oblanceolate to striped lanceolate, acute at the apex; leaves gray strigose; flowers small, light blue, sessile, flower solitary or several terminal racemes, bracts striped; calyx striped quinquelobed, blue, the throat of corolla has 5 appendants; stamens 5, stigmas capitellate; corolla obtuse, quinquelobed; corolla tubular or funnel-shaped; flower disc existing; ovary superior, consisting of two carpels, each carpel has two ovules. Fruits 4 top-shaped nutlets, tuberculate and 5～100 toothlets.

〔Habitat〕

Mountain top, stony hillsides grasslands, gravel tunnels or roadsides.

〔Use〕

Medicinal use, clearing heat and poison.

① 植株
② 花
③ 叶

北齿缘草　*Eritrichium borealisinense* Kitag.

〔特征〕

多年生草本，高15～40cm。根粗壮，直径可达1cm。茎数条，常形成不大的密簇。基生叶倒披针形，长3～6（8）cm，宽（3）4～8mm，先端急尖，基部楔形，两面密被长短粗细不同的糙伏毛；茎生叶倒披针形或披针形，长1.5～3cm，宽（3）4～8mm，缘毛明显或不明显。花序分枝2或3（4）个，长1～2cm，果期长2～5（10）cm，分枝有数至十数朵花，生苞腋外；苞片线状披针形，长3～5mm；花梗较粗壮，长2～5（7）mm，直立或稍斜伸，生白伏毛；花萼裂片长圆状披针形至长圆状线形，长3.5～4mm，宽1～1.5mm，先端渐尖至急尖，两面生糙伏毛，花期直立，果期多斜展；花冠蓝色，钟状辐形，筒长1.2～1.5mm，裂片倒卵形或近圆形，长3～3.5mm，附属物半月形至矮梯形，伸出喉外；花药长圆形。小坚果近陀螺状，除棱缘的刺外，长2～2.5mm，宽1mm，背面卵形或宽卵形，微凸，密布小疣突和刚毛，中肋明显，腹面粗糙，有龙骨状突起，着生面三角形，位于腹面中部或中下部，边缘有三角形锚状刺，刺具微毛基部连合或近分离。花果期7—9月。

〔生境〕

生于山坡草地、石缝、灌丛和石质干山坡。

〔用途〕

药用，清温解热。

〔Features〕

Perennial herb, 15 ~ 40cm tall. Roots stout, up to 1cm in diameter. Stems several, usually forming into a dense cluster. Basal leaves oblanceolate, 3 ~ 6(8)mm long, (3)4 ~ 8mm wide, acute at the apex, cuneate at the base, densely strigose of different sizes on both sides; cauline leaves oblanceolate or lanceolate, 1.5 ~ 3cm long, (3)4 ~ 8mm wide, tricholoma conspicuous or inconspicuous. Inflorescence 2 or 3(4) branched, 1 ~ 2cm long, 2 ~ 5(10)cm long in the fruiting period, several to dozens of flowers grow outside of axils; bracts linear-lanceolate, 3 ~ 5mm long; pedicels stout, 2 ~ 5(7)mm long, erect or slightly oblique, white adpressed hairs; calyx lobes oblong-elliptic lanceolate to oblong-elliptic linear, 3.5 ~ 4mm long, 1 ~ 1.5mm wide, acuminate to acute at the apex, strigose on both sides, erect when flowering, mostly obliquely unfolding in the fruiting period; corolla blue, campaniform, tube 1.2 ~ 1.5mm long, lobes obovate or suborbicular, 3 ~ 3.5mm long, appendant semilunar to ladder-shaped, extended out of throat; anthers long elliptic. Nutlets nearly turbinate, except thorns of carinal edge, 2 ~ 2.5mm long, 1mm wide, ovoid or broadly ovoid on the back, convex, densely protuberances and bristles, midribs conspicuous, ventral side coarse, with carina, areole triangular, at middle or lower middle part of ventral side, triangular anchor-shaped thorns at the margin, thorns connective or nearly separated at the base. Flowering and fruiting July to September.

〔Habitat〕

Hillside meadows, rock tunnels, bushwood and stony dry hillside.

〔Use〕

Medicinal use, antipyretic.

① 植株
② 花
③ 茎、叶、果实、花
④ 小花
⑤ 成熟枝条与荚果

鹤虱属　*Lappula* V. Wolf

鹤　虱　*Lappula myosotis* V. Wolf.

〔特征〕

一年生或二年生草本。茎直立，高20～35cm，中部以上多分枝，全株（茎、叶、苞片、花梗、花萼）均密被白色细刚毛。基生叶矩圆状匙形，全缘，先端钝，基部渐狭下延，长达7cm（包括叶柄在内），宽3～9mm；基生叶较短而狭，披针形或条形，长3～4cm，宽15～40mm，扁平或沿中肋纵折，先端尖，基部渐狭，无叶柄。花序在花期较短，果期则伸长，长5～12cm；花梗果期伸长，长约2cm，直立；花萼5深裂至基部，裂片条形，锐尖，花期长2mm，果期增大呈狭披针形，长约3mm，宽0.7mm，星状开展或反折；花冠浅蓝色，漏斗状至钟状，长约3mm，裂片矩圆形，长1.2m，宽1.1mm，喉部具5矩圆形附属物；花药矩圆形，长0.5mm，宽0.3mm；花柱长0.5mm，柱头扁球形。小坚果卵形，长3～3.5mm，基部宽0.8mm，背面狭卵形或矩圆状披针形，通常有颗粒状瘤凸，稀平滑或沿中线龙骨状突起上有小棘突，背面边缘有2行近等长的锚状刺，内行刺长1.5～2mm，基部不互相汇合，外行刺较内行刺稍短或近等长，通常直立；小坚果侧面通常具皱纹或小瘤状凸起，花柱高出小坚果但不超出小坚果上方之刺。花果期6—8月。

〔生境〕

生于合谷草甸、山地草甸、路旁。

〔用途〕

药用，杀虫、具有清热解毒、健脾和胃功效。

〔Features〕

Annual or biennial herb. Stems erect, 20 ~ 35cm tall, multi-branched above the middle part, the whole plant (stems, leaves, bracts, peduncles, calyx) is densely covered with white finely bristles. Basal leaves rounded-rectangular spathulate, the margin entire, obtuse at the apex, tapering decurrent at the base, up to 7cm long (including petioles), 3 ~ 9mm wide; basal leaves short and narrow, lanceolate or striped, 3 ~ 4cm long, 15 ~ 40mm wide, flat or longitudinal fracture along the keel, acute at the apex, tapering at the base, no petioles. Inflorescence short during the flowering phase, and elongate during fruiting period, 5 ~ 12cm long; peduncles elongate during fruiting period, about 2cm long, erect; calyx 5-parted to the base, lobes striped, acuminate, 2mm-long during flowering phase, accrescent and narrowly lanceolate during fruiting period, about 3mm long, 0.7mm wide, starlike squarrose or reflexed; corolla light blue, funnel-shaped to campaniform, about 3mm long, lobes oblong, 1.2m long, 1.1mm wide, the throat has 5 oblong appendants; anthers oblong, 0.5mm long, 0.3mm wide; stylus 0.5mm long, stigmas flat spherical. Nutlets ovate, 3 ~ 3.5mm long, 0.8mm wide at the base, narrowly ovate or rounded-rectangular lanceolate on the back, usually granular hump, rarely smooth or keels protruding spinose, 2 lines of equilong anchoreaform thorns on the back of the margin, inner thorns 1.5 ~ 2mm long, disjoint at the base, outer thorns slightly shorter than or approximately equal in length to inner thorns, usually erect; the side face of nutlets usually rugose or papillose, stylus above nutlets but not the thorns of nutlets. Flowering and fruiting June to August.

〔Habitat〕

Valley meadows, mountain meadows and roadsides .

〔Use〕

Medicinal use, pesticidal, clearing heat and removing toxicity, and strengthening the spleen and stomach.

① 植株 ② 株丛 ③ 叶片 ④ 花 ⑤ 叶

唇形科 Labiatae

青兰属 *Dracocephalum* Linn. Nom. Conserve.

香青兰 *Dracocephalum moldavica* L.

〔特征〕

一年生草本，全株密被短毛。茎方形，高20 ~ 40cm茎基部分枝斜升。叶对生，有短柄，叶片卵状披针形，边缘具钝圆齿，背面有腺点，两面有短细毛。花生于茎或分枝上部叶腋内，轮伞花序，每3 ~ 6朵一轮，秋季开花；苞片矩圆状楔形，下部有2 ~ 3对刺；花冠唇形，淡蓝紫色，长约2cm，上唇微凹，下唇中裂片两裂；花萼有15条脉，上萼有3齿，下萼有2齿；雄蕊4，花柱头两裂，小坚果长圆形。

〔生境〕

常生干燥地，多见于田边、路旁、荒地、草原等地。

〔用途〕

药用，具有清热燥湿、凉血功效。

〔Features〕

Annual herb, the whole plant densely short-pilose. Stems square, 20 ~ 40cm tall, the base of stems branched, oblique ascending. Leaves opposite, short petioles, leaf-blades ovate-lanceolate, obtuse crenate at the margin, glandular dots on the back, short tiny hairs on both surfaces. Flowers grow in upper axils of stems or branches, verticillaster, each 3 ~ 6 flowers a whorl, blooming in autumn; bracts rounded-rectangular cuneate, 2 ~ 3 pairs of thorns below; corolla labiate, light blue purple, about 2cm long, labrum retuse, labium lobed bifid; calyx 15-veined, upper calyx 3 teeth, lower calyx 2 teeth; stamens 4, stigma bifid, nutlets long elliptic.

〔Habitat〕

Dry lands, commonly seen in fields, roadsides, wastelands and grasslands, etc.

〔Use〕

Medicinal use, clearing heat, eliminating dampness and blood cooling.

① 株丛 ② 花序 ③ 茎、叶 ④ 果实与花蕾 ⑤ 花蕾

黄芩属 *Scutellaria* Linn.

黄　芩 *Scutellaria baicalensis* Georgi

〔特征〕

多年生直立草本，高15～120cm。根肥大，圆柱形，长50cm，皮黑褐色，中心黄色。茎四棱，基部分枝丛生，基部木质化，叶对生，披针形至条状披针形，下面密被下陷的腺点，全缘。夏季开花，花唇形，蓝色，聚生成顶生总状花序，花偏向一侧；花萼两侧唇形，花后增大；苞片叶状，卵状披针形；花萼紫绿色，密被毛；花冠紫色或蓝色，花冠筒部长，顶端两唇形；雄蕊4，二强；花药被毛；花柱纤细，两裂；子房4深裂。小坚果4个，近球形，黑色，包裹在宿存花被内。

〔生境〕

生于平原草地、山坡草地、灌丛中、向阳草坡及休荒地上。

〔用途〕

药用，具有抗菌、降压作用。

〔Features〕

Perennial erect herb,15～120cm tall. Roots hypertrophic, cylindrical, 50cm long, root bark black brown, the center yellow. Stems tetragonous, branches fasciculate at the base, lignified, leaves opposite, lanceolate to striped-lanceolate, dense sunken glandular dots beneath, the margin entire. Blooming in summer, flowers labiate, blue, arranged into terminal racemes, flowers inclined to one side; both sides of calyx labiate, accrescent post floral; bracts lobate, ovate-lanceolate; calyx purple green, covered with dense hairs; corolla purple or blue, corolla tube long, apex labiate; stamens 4, didynamous; anthers pilose; stylus slender, bifid; ovary 4-parted. Nutlets 4, subsphaeroidal, black, wrapped in persistent perianth.

〔Habitat〕

Plain grasslands, hillside grasslands, bushwoods, sunny grass slopes and wastelands.

〔Use〕

Medicinal use, antibacterial and antihypertensive effect.

① 植株　② 花头部　③ 花、叶　④ 花、叶

并头黄芩　*Scutellaria scordifolia* Fisch. ex Schrank

〔特征〕

多年生草本，高10～30cm。根茎细长，淡黄白色。茎直立或斜生，四棱形，沿棱疏被微柔毛或近几无毛，单生或分支。叶三角状披针形、条状披针形或披针形，长1.7～3.3cm，宽3～11cm，先端钝或稀微尖，基部圆形、浅心形、心形乃至截形，边缘具疏锯齿或全缘；上面被短柔毛或无毛，下面沿脉被微柔毛，具多数凹腺点；具短叶柄或几无柄。花单生于茎上部叶腋内，偏向一侧；花梗长3～4mm，近基部有1对长约1mm的针状小苞片；花萼疏被短柔毛，果后花萼长达4～5mm，盾片高2mm；花冠蓝色或蓝紫色，长1.8～2.4cm，外面被短柔毛，冠筒基部浅囊状膝曲，上唇盔状，内凹，下唇3裂；子房裂片等大，黄色，花柱细长，先端锐尖，微裂。小坚果近圆形或椭圆形，长0.9～1mm，宽0.6mm，褐色，具瘤状突起，腹部中间具果脐，隆起。花期6—8月；果期8—9月。

〔生境〕

生于河滩草甸、山地草甸、山地林缘、林下、撂荒地、路旁。

〔用途〕

药用。

〔Features〕

Perennial herb, 10 ~ 30cm tall. Rhizomes elongated, yellowish white. Stems erect or oblique, four prismatic, sparsely puberulent or glabrescent along the ridges, solitary or branched. Leaves triangular lanceolate, striped lanceolate or lanceolate, 1.7 ~ 3.3cm long, 3 ~ 11cm wide, obtuse or sparsely acutate at the apex, rounded,shallowly cordate at the base, heart-shaped to truncate, sparsely serrulate at the margin or the margin entire; short pubescenct at the top or glabrous, puberulent beneath along the eidges, with numerous concave glandular dots; short petioles or nearly impedicellate. Flowers solitary within axil at the top, inclined to one side; peduncles 3 ~ 4mm long, a pair of 1mm-long needle-like bractlets near the base; calyx sparsely pubescent, calyx up to 4 ~ 5mm long after fruiting, scutellum 2mm high; corolla blue or blue purple, 1.8 ~ 2.4cm long, short pubescent on the outside, corolla tube shallowly saccate geniculate at the base, labrum cuculliform, indent, labium 3-lobed; lobes of ovary isometrical, yellow, stylus elongated, acuminate at the apex, lobed. Nutlets uborbicular or oval, 0.9 ~ 1mm long, 0.6mm wide, brown, tuberculate, areola in the middle of abdomen, ridgy. Flowering June to August; and fruiting August to September.

〔Habitat〕

Beachlands, meadows, mountain meadows, mountain lands, forest edges, wastelands and roadsides.

〔Use〕

Medicinal use.

① 植株
② 花
③ 株丛

糙苏属 *Phlomis* Linn.

串铃草 *Phlomis mongolica* Turcz.

〔特征〕

多年生草本。根木质化，较粗。茎直立，四棱，高20～70cm，被糙毛。叶对生，长三角状卵形或三角状披针形，先端钝，基部心形，边缘具钝锯齿，厚而糙，上面有毛，下面密生星状毛，叶柄长10cm左右。轮伞花序生于茎、枝上部叶腋内；苞片条状形，边缘具长睫毛；花萼管状，长1.4cm，先端5裂，沿脉有糙毛；花冠紫红色，长2.2cm，唇形，上唇盾状，里面密被白色长绵毛，下唇三裂，稍外卷，外面有毛；雄蕊4，二强，花丝基部有距状物；花柱一枚，两裂。小坚果卵形，顶端有毛。

〔生境〕

生于干燥草坡、山坡草地。

〔用途〕

药用。

〔Features〕

Perennial herb, roots lignified, thick. Stems erect, tetragonous, 20～70cm tall, sparsely strigose. Leaves opposite, long triangular or triangular-lanceolate, acute at the apex, heart-shaped at the base, crenate at the margin, thick and coarse, hairy at the top, densely stellate hairs beneath, petioles about 10cm long. Verticillaster born on axil of stems and branches; bracts striped, ciliiform at the margin; calyx tubular, 1.4cm long, apex 5-lobed, strigose along the veins; corolla aubergine, 2.2cm long, labiate, labrum peltate, white long pubescent on the inside; labium 3-lobed, slightly revolute, pilose on the outside; stamens 4, didynamous, the base of filament calcariform, stylus 1, bifid. Nutlets ovate, pilose at the apex.

〔Habitat〕

Dry grass slopes and hillside meadows.

〔Use〕

Medicinal use.

① 植株 ② 花序 ③ 叶片

裂叶荆芥属 *Schizonepeta* Briq.

多裂叶荆介 *Schizonepeta multifida* (L.) Briq.

〔特征〕

一年生草本，有强烈香气。茎直立，具四棱，上部少分枝，密被逆白毛。叶对生，阔卵形，茎基部叶羽状5裂，中部叶3裂，也有全缘；裂片卵形或卵状披针形；叶两面被短白毛，背面有黄色腺毛，叶柄短；轮生花序密集成顶生假穗状花序；苞片卵形，先端骤尖，淡蓝紫色；花萼钟状，5齿裂，具15脉，淡蓝色，背面密被毛，混生细纹状腺点；花冠唇形，蓝紫色；上唇2裂，下唇3裂；雌蕊1枚，花柱伸出，雄蕊4，二强。小坚果椭圆形，纵肋3，有小点。

〔生境〕

生于草甸、路旁、山坡草地、平原草地上。

〔用途〕

药用，具有发汗解表功能。

〔Features〕

Annual herb, strong aromatic. Stems erect, tetragonous, a few branches at the top, densely retrorsely white hairs. Leaves opposite, broadly ovate, leaves at basal part of stem pinnate and 5-lobed, the middle part 3-lobed, occasionally the margin entire; lobes ovate or ovate-lanceolate; short white pilose on both surfaces of leaves, yellow glandular hairs on the back, petioles short; verticillate inflorescence clustered into terminal spike; bracts ovate, cuspidate at the apex,light purple blue; calyx campaniform, 5 dentate fissure, 15-veined, light blue, covered with dense hairs on the back, mixed pinstripedes glandular dots; corolla labiate, blue purple; labrum bifid, labium 3-lobed; pistil 1, stylus extended, stamens 4, didynamous. Nutlets oval, longitudinal ribs 3, punctate.

〔Habitat〕

Meadows, roadsides, hillside grasslands, and plain grasslands.

〔Use〕

Medicinal use, inducing perspiration and relieving exterior syndrome.

③ 植株 ② 叶 ③ 花序

益母草属 *Leonurus* Linn.

细叶益母草 *Leonurus sibiricus* L.

〔特征〕

一年生或二年生草本。茎直立，高20～80cm，茎四棱，被白色逆毛。叶对生，基生叶具长柄，近圆形，5～9浅裂，边缘具钝锯齿，两面密被柔毛；茎中生叶具短柄，卵形，掌状二全裂，茎上生叶二全裂，裂片再二全裂。夏季开花，花唇形，淡红色或红色，轮生在茎上部的叶腋内，花多数；苞片锥形或针形，无毛；花萼钟形，密被柔毛，具刺毛状5裂片，有5脉。雄蕊4，两个雄蕊退化，花柱头两裂。小坚果4，三角形，黑褐色。

〔生境〕

生于山地丘陵地带、平原、荒地、草原和山地草原。

〔用途〕

药用，具有活血、祛瘀、调经功效。

〔Features〕

Annual or biennial herb. Stems erect, 20～80cm tall, tetragonous, white retrorse pilose. Leaves opposite, basal leaves long petioles, suborbicular, 5～9-lobed, crenante at the margin, densely pubescent on both surfaces; leaves grow on stems have short petioles, ovate, palmate pinnatisect, leaves pinnatisect, lobes pinnatisect. Blooming in summer, flowers labiate, light red or red, verticillate in the axil at the top of stems, flowers numerous; bracts subulate or aciculiform, glabrous; calyx campaniform, densely pubescent, setaceous 5 lobes, 5-veined. Stamens 4, 2 stamens vestigial, stylus bifid. Nutlets 4, triangle, black brown.

〔Habitat〕

Mountain lands hills, plains, wastelands, grasslands and mountain grasslands.

〔Use〕

Medicinal use, activating blood circulation, removing stasis and regulating menstruation.

① 花、叶 ② 叶

百里香属 *Thymus* Linn.

百里香 *Thymus mongolicus* Ronn.

〔特征〕

多年生灌木状芳香草本，茎带红色，匍匐地上或半斜升，茎和花枝高2～10cm，密丛生，基部木质化，被柔毛。叶小，对生；卵形或长椭圆状披针形，顶端钝，基部渐尖，全缘；有不明显2～3对叶脉；具极短柄，全株芳香。夏季开花，轮伞花序密集成头状，花萼两唇形，五裂，边缘有睫毛；花冠两唇形，紫红色，上唇宽卵圆形，顶端两浅裂，下唇三裂；雄蕊4，二强，外伸；柱头两裂，花柱一枚。小坚果卵形，光滑。

〔生境〕

生于沙砾质土壤、向阳坡草地、草原。

〔用途〕

药用；可提取芳香料。

〔Features〕

Perennial fruticose fragrant herb, stems reddish, decumbent or semi-obliquely ascending, stems and sprays 2～10cm tall, densely fasciculate, lignified at the base, pubescent. Leaves small, opposite; ovate or long elliptical lanceolate, obtuse at the apex, acuminate at the base, the margin entire; inconspicuous 2～3 pairs of veins; very short petioles, the whole plant fragrant. Blooming in summer, verticillaster clustered into capitellate, calyx labiate, quinquelobed, ciliiform at the margin; corolla labiate, aubergine, labrum broadly oval, apex 2-lobed, labium trilobate; stamens 4, didynamous, lituate; stigmas bifid, stylus 1. Nutlets ovate, glabrous.

〔Habitat〕

Gravel soil, sunny lands and grasslands.

〔Use〕

Medicinal use; flavor extracts.

① 植株 ② 叶

香薷属 *Elsholtzia* Willd

香　薷 *Nepeta cataria* Linn

〔特征〕

多年生植物。茎坚强，基部木质化，多分枝，高40～150cm，基部近四棱形，上部钝四棱形，具浅槽，被白色短柔毛。叶卵状至三角状心脏形，长2.5～7cm，宽2.1～4.7cm，先端钝至锐尖，基部心形至截形，边缘具粗圆齿或牙齿，草质，上面黄绿色，被极短硬毛，下面略发白，被短柔毛但脉上较密，侧脉3～4对，斜上升，在上面微凹陷，下面隆起；叶柄长0.7～3cm，细弱。花序为聚伞状，下部的腋生，上部的组成连续或间断的、较疏松或极密集的顶生分枝圆锥花序，聚伞花序呈二歧状分枝；苞叶叶状，或上部的变小而呈披针状，苞片、小苞片钻形，细小。花萼花时管状，长约6mm，径1.2mm，外被白色短柔毛，内面仅萼齿被疏硬毛，齿锥形，长1.5～2mm，后齿较长，花后花萼增大成瓮状，纵肋十分清晰。花冠白色，下唇有紫点，外被白色柔毛，内面在喉部被短柔毛，长约7.5mm，冠筒极细，径约0.3mm，自萼筒内骤然扩展成宽喉，冠檐二唇形，上唇短，长约2mm，宽约3mm，先端具浅凹，下唇3裂，中裂片近圆形，长约3mm，宽约4mm，基部心形，边缘具粗牙齿，侧裂片圆裂片状。雄蕊内藏，花丝扁平，无毛。花柱线形，先端2等裂。花盘杯状，裂片明显。子房无毛。小坚果卵形，几三棱状，灰褐色，长约1.7mm，径约1mm。花期7—9月，果期9—10月。

〔生境〕

生于宅旁或灌丛中。

〔用途〕

药用。

〔Features〕

Perennial plant. Stems stout, lignified at the base, multi-branches, 40 ~ 150cm tall, nearly tetragonous at the base, tetragonous at the top, shallowly fluted, white and short pubescent. Leaves ovoid to triangular heart-shaped, 2.5 ~ 7cm long, 2.1 ~ 4.7cm wide, obtuse to acute at the apex, cordate to truncate at the base, coarsely crenate or dentate at the margin, herbaceous, yellowish green at the top, hispidulous, slightly whitish beneath, short pubescent and dense along veins, lateral veins 3 ~ 4-paired, obliquely ascending, retuse at the top, ridgy beneath; petioles 0.7 ~ 3cm long, thin and delicate. Inflorescene cymose, axillary at lower part, continuous or discontinuous, sparsely or densely terminal panicles at the top, cymes dichotomy branched; bracts leafy, diminish and lanceate at upper part, bracts and bractlets subulate and thin. Calyx tubular when flowering, about 6mm long, 1.2mm in diameter, white and short pubescent outside, calyx teeth sparse strigose inside, teeth tapering, 1.5 ~ 2mm long, rear tine longer, calyx accrescent to gyalectiform when flowering, longitudinal ribs very conspicuous. Corolla white, labium with purple dots, white pubescent on the outside, the throat short pubescent on the inside, about 7.5mm long, corolla tube very thin, about 0.3mm in diameter, expanding to wide throat from calyx tube, limbs bilabiate, labrum short, about 2mm long, about 3mm wide, retuse at the apex, labium 3-lobed, central lobes suborbicular, about 3mm long, about 4mm wide, cordate at the base, crenate at the margin, lateral lobes rounded. Stamens concealed, filaments flat, glabrous. Stylus linear, 2-lobed at the apex. Flower disc cyathiform, lobes conspicuous. Ovary glabrous. Nutlets ovoid, nearly trigonous, grey brown, about 1.7mm long, about 1mm in diameter. Flowering July to September, and fruiting September to October.

〔Habitat〕

Next to residential houses or bushwood.

〔Use〕

Medicinal use.

① 植株 ② 花 ③ 叶 ④ 果实

茄 科 Solanaceae

天仙子属 *Hyoscyamus* L.

天仙子 *Hyoscyamus niger* L.

〔特征〕

一年生或二年生有毒草本，高30～70cm，根纺锤状粗壮。全株有黏性腺毛，并有特殊臭气，味苦。叶互生，矩圆形，长7～25cm，宽2～6cm，边缘有疏齿牙或羽状分裂；茎下部叶有柄，上部叶无柄并叶基部抱茎。夏季开花，花漏斗状，土黄色，有紫色网状脉纹；穗状花序集于茎或枝端，花单生叶腋，花多数；花萼坛状，长10～15mm，密被腺毛，裂片5枚，三角形，顶端针刺状；雄蕊5，花丝深紫色；雌蕊1枚，长约2cm；子房球形，二室胚珠多数；花柱细长，伸出花冠外，柱头圆柱状。蒴果球形，包藏于增大的宿萼内，盖状开裂，种子扁平，多数，淡黄褐色。

〔生境〕

生于宅旁、路边、荒地、河边沙质地。

〔用途〕

药用，具有镇痉，止痛的功效。

〔Features〕

Annual or biennial poisonous herb, 30～70cm tall, roots cambiform stout. The whole plant has viscous glandular hairs, with special bad smell, bitter in taste. Leaves alternate, oblong, 7～25cm long, 2～6cm wide, sparsely dentate or pinnately divided at the margin; leaves petiolate at lower part of stems, leaves impedicellate at the top and amplexicaul at the base. Bloom in summer, flowers funnel-shaped, earthly yellow, with purple reticulate veinings; spike centered at stems or the ends of the branches; flowers solitary in the axils, flowers numerous; calyx urceolus, 10～15mm long, densely glandular hairs, lobes 5, triangle, apex spinous; stamens 5, filaments deep purple; pistil 1, about 2cm long; ovary spherical, dithecal ovules numerous; stylus elongated, stretching out of corolla, stigmas cylindrical. Capsules spherical, wrapped in accrescent persistent calyx, hooded cracked, seeds flat, numerous, light yellowish-brown.

〔Habitat〕

Next to residential houses,roadsides,wastelands ,river bankssandy lands.

〔Use〕

Medicinal use, relieving muscular spasm and pain.

① 植株
② 花序

玄参科 Scrophulariaceae

婆婆纳属 *Veronica* L.

白婆婆纳 *Veronica incana* Linn

〔特征〕

植株全体密被白色绵毛，呈白色，仅叶上面较稀而呈灰绿色。茎数支丛生，直立或上升，不分枝，高15～40cm。叶对生，上部的有时互生，下部的叶片矩圆形至椭圆形，上部的常为宽条形，长1.5～5cm，宽0.3~1.5cm，顶端钝至急尖，基部楔状渐窄，下部的叶具长达2cm的柄，上部的近无柄，叶缘具圆钝齿或全缘。花序长穗状；花梗极短；花萼长约2mm，花冠蓝色，蓝紫色或白色，长5～7mm，筒长1.5～2mm，裂片常反折，圆形，卵圆形至卵形；雄蕊略伸出；子房及花柱下部被多细胞腺毛。蒴果长略超过花萼，被毛。花期6—8月。

〔生境〕

生草原及沙丘上。

〔用途〕

药用，清热消肿、凉血止血。

〔Features〕

The whole plant is densely covered with white lanate, white, only sparse leaves on the top are grey green. Several stems fasciculate, erect or ascending, eramose, 15～40cm tall. Leaves opposite, upper leaves occasionally alternate, lower leaf-blades rounded-rectangular to ellipse, upper part usually broadly striped, 1.5～5cm long, 0.3～1.5cm wide, obtuse to acute at the apex, sphenoid and gradually narrow at the base, lower leaves with 2cm-long petioles, upper part nearly impedicellate, leaf margins crenate or the margin entire. Inflorescences spicate; pedicels very short; calyx about 2mm long, corolla blue, blue purple or white, 5～7mm long, tubes 1.5～2mm long, lobes usually reflexed, rounded, oval to ovoid; stamens slightly extended; lower part of ovary and stylus covered with glandular hairs. Capsules slightly longer than calyx, hairy. Flowering June to August.

〔Habitat〕

Grasslands and sand dune.

〔Use〕

Medicinal use, clearing heat, mitigating edema, cooling blood and hemostatic.

① 植株 ② 花冠 ③ 叶

马先蒿属 *Pedicularis* Linn.

红纹马先蒿 *Pedicularis striata* Pall.

〔特征〕

多年生草本，高20～40cm，根粗大。茎基部丛生，直立，无分枝，全株被柔毛。叶线生，4～8cm长，羽状深裂或全裂，小裂片条形，边缘具细锯齿，顶端稍尖或钝；茎基部叶具柄，上部叶无柄；基部叶单生或双生。穗状花序顶生，花多数，密集；苞片叶状，矩圆形，全缘或具细锯齿；萼筒圆筒形，疏生毛，长1cm，先端四列；花冠黄色，有紫褐色冠脉，长2.5～3cm，唇形，上唇钩状弯曲或镰刀状弯曲，通常喙状突起，下唇先端钝，小裂片3枚；雄蕊4，不外伸；子房上位，两室，每室有胚珠多数。蒴果椭圆形，顶端尖，深褐色，开裂，包于宿存萼内。种子多数，具网状孔纹，灰黑褐色。

〔生境〕

生于山坡草地、山谷、草原、林缘、灌丛中。

〔用途〕

药用，主治蛇虫咬伤、水肿、耳鸣、口干舌燥、痈肿等。

〔Features〕

Perennial herb, 20～40cm tall, roots thick and big. Stems fasciculate at the base, erect, eramose, the whole plant pubescent. Leaves linear, 4～8cm long, pinnatiparted or pinnatisect, lobelets striped, serrate at the margin, slightly acute or obtuse at the apex; leaves at basal part of stems petiolate, upper leaves impedicellate; leaves solitary or binate at the base. Spica terminal, flowers numerous, dense; bracts leafy, rounded-rectangular, the margin entire or serrate; calyx tube cylindrical, sparse pilose, 1cm long, four-row at the apex; corolla yellow, puce coronary artery, 2.5～3cm long, labiate, labrum uncinated or falciform curved, usually coronoid, labium obtuse at the apex, lobelets 3; stamens 4, not lituate; ovary superior, 2 locules, each locule has numerous ovules. Capsules ellipse, acute at the apex, dark brown, cracked, wrapped in the persistent calyx. Seeds numerous, with reticulate stripes, black brown.

〔Habitat〕

Hillside meadows, valleys, grasslands, forest edges and bushwoods.

〔Use〕

Medicinal use, treating bites caused by insects or snakes, edema, tinnitus, mouth parched and tongue scorched and swollen welling-abscess, etc.

① 植株 ② 花 ③ 花

柳穿鱼属 *Linaria* Mill.

柳穿鱼 *Linaria vulgaris* Mill.

〔特征〕

多年生草本，高20～70cm，主根细长黄白色。茎直立，分枝或不分枝，灰白色。叶互生，无柄，条形至条状披针形，先端尖，全缘，基部楔形，长3cm，无毛。总状花序顶生，各部被腺毛，夏秋开花；花有梗，淡黄色；花萼5深裂，裂片条状披针形，基部合并；花冠唇形，喉部密被毛，基部成长距，上唇一裂，下唇两裂，开展；喉部顶端橘红色，被长毛；雄蕊4，二强，不外伸；子房上位，二室，胚珠多数；花柱细，生于子房上位，有小裂口。蒴果卵形或卵圆形，顶端六瓣裂，开裂。

〔生境〕

生于山坡草地、草原、沙质地、路旁、河边、沙砾质草原及沙地。

〔用途〕

药用，具有清热解毒功效。

〔Features〕

Perennial herb, 20～70cm tall, taproots elongated yellowish white. Stems erect, branched or unbranched, grey white. Leaves alternate, impedicellate, striped to striped-lanceolate, acute at the apex, the margin entire, cuneate at the base, 3cm long, glabrous. Racemes terminal, all parts clothed with glandular hairs, blooming in summer and autumn; flowers pediculate, light yellow; calyx 5-parted, lobes striped-lanceolate, merged at the base; corolla labiate, throat covered with dense hairs, long spaced at the base, labrum 1 lobed, labium bifid, squarrose; the top of throat orange, long-pilose; stamens 4, didynamous, not lituate; ovary superior, 2 locules, ovules numerous; stylus slender, grow on ovary superior, with small cracks. Capsules ovate or oval, six-petaled at the apex, cracked.

〔Habitat〕

Hillside meadows, grasslands, sandy lands, roadsides, river banks, gravel grasslands and sand lands.

〔Use〕

Medicinal use，clearing heat and remoring to xicity.

① 植株 ② 花 ③ 花朵

芯芭属 *Cymbaria* Linn.

达乌里芯芭 *Cymbaria dahurica* L.

〔特征〕

多年生草本，高5～20cm，根粗，红色。茎直立，基部多分枝丛生；全株密被银白色绢毛。叶对生，无柄，茎上部叶互生，顶端急尖，上面淡绿色，背面灰白色；茎基部叶小，密集，椭圆形或椭圆状披针形；茎中部和上部叶条形或条状披针形，全缘。花茎上部叶腋内单生，有花梗，基部具小苞片两枚；花萼5裂，花冠很大，鲜金黄色，两唇形，上唇盾形，有毛，下唇三裂；雄蕊4，二强；花柱细。蒴果长卵形，种子多数。

〔生境〕

生于草原、山坡草地、沙质地上。

〔用途〕

药用。

〔Features〕

Perennial herb, 5～20cm tall, roots thick, red. Stems erect, multi-branched fasciculate at the base; the whole plant is densely covered with silver-white sericeous. Leaves opposite, impedicellate, leaves alternate at the top of stems, abruptly acute at the apex, light green at the top, grey white on the back; small leaves at basal part of stem, dense, oval or elliptical-lanceolate; leaves striped or striped-lanceolate at middle-upper part of stems, the margin entire. Rachises solitary in axils, with peduncles, 2 bractlets at the base; calyx 5-lobed, corolla very large, golden yellow, bilabiate, labrum peltate, hairy, labium trilobate; stamens 4, didynamous; stylus slender. Capsules long ovate, seeds numerous.

〔Habitat〕

Grasslands, hillside meadows and sandy lands.

〔Use〕

Medicinal use.

① 植株 ② 小穗 ③ 叶

车前科 Plantaginaceae

车前属 *Plantago* L.

平车前 *Plantago depressa* willd.

〔特征〕

一年生或二年生草本，主根发达。叶基生，广卵形或长椭圆状卵形，有长柄，基部狭楔形，顶端钝尖，边缘具不规则锯齿。花葶1～10，出自叶丛间，高10～30cm，穗状花序由叶丛中央生出，下部花较疏散，上部花较密；夏秋开花；苞片绿色，边缘白色；花萼裂片4枚，基部稍合生；花被筒状，裂片4枚，边缘干膜质，裂片先端浅凹；雄蕊4；雌蕊1；子房一位，二室；花柱纤细，柱头密被毛；蒴果椭圆形，盖裂，种子4～5粒，矩圆珠笔形，较小，黑棕色，光滑。花、果期6—10月。

〔生境〕

生于草原、路旁、沙质湿土壤、宅旁等。

〔用途〕

药用，具有利水通淋、清热明目功效。

〔Features〕

Annual or biennial herb, taproots developed. Leaves basal, broadly ovate or long elliptical ovate, with long petioles, narrowly cuneate at the base, obtuse to sharp at the apex, irregular serrate at the margin. Peduncles 1～10, from leafage, 10～30cm tall, spikes grow from center of leafage, lower part flowers scattered, flowers dense at the top; blooming in summer and autumn; bracts green, white at the margin; calyx lobes 4, slightly concrescent at the base; perianth tubular, lobes 4, scarious at the margin, lobes retuse at the apex; stamens 4; pistil 1; ovary 1, 2 locules; stylus slender, stigmas covered with dense hairs; capsules oval, dehiscentbylid, seeds 4～5, rounded-rectangular ballpoint pen-shaped, small, balck brown, glabrous. Flowering and fruiting June to October.

〔Habitat〕

Grasslands, roadsides, sandy wet soil, next to residential houses, etc.

〔Use〕

Medicinal use, promoting diuresis and relieving stranguria, clearing heat and improving eyesight.

① 植株 ② 叶与茎 ③ 花序 ④ 花

茜草科 Rubiaceae

拉拉藤属 *Galium* Linn.

蓬子菜 *Galium verum* L.

〔特征〕

灌木状多年生草本。高30 ~ 80cm，茎直立，基部稍木质化，具4纵棱，无毛，幼时被柔毛。通常每节轮生6 ~ 10叶，长1.5 ~ 4cm，宽1 ~ 1.5mm，狭条或条形，中脉一条，边缘有细锯齿或稍反卷，背面被柔毛；无叶柄，聚伞圆锥花序顶生或上部叶腋生，花梗被细灰白毛；花小，两性；花萼卵形或球形；花冠淡黄色或黄色，或黄白色，裂片矩圆状披针形，4枚，外伸；子房下位，两室，每室有胚珠一枚，花柱2裂。种子两两双生，扁球形，无毛，黄褐色。

〔生境〕

生于潮湿草原、草甸草原、山坡、沟谷、林边湿地上。

〔用途〕

药用。

〔Features〕

Fruticose perennial herb. 30 ~ 80cm tall, stems erect, slightly lignified at the base, with 4 longitudinal ridges, glabrous, pubescent when young. Usually each internode verticillate 6 ~ 10 leaves, 1.5 ~ 4cm long, 1 ~ 1.5mm wide, narrowly striped or striped, midrib 1, serrulate or slightly deflexed at the margin, pubescent on the back; no petioles, cymes terminal or leaves axillary at the top, peduncles finely incanous; flowers small, bisexual; calyx ovate or spherical; corolla light yellow or yellow, or yellowish white, lobes 4 rounded-rectangular lanceolate, lituate; ovary inferior, 2 locules, each locule has 1 ovule, stylus bifid. Seeds binate, flat spherical, glabrous, yellowish-brown.

〔Habitat〕

Wet grasslands, meadow steppes, hillsides, cheuches, forest edges and wetlands.

〔Use〕

Medicinal use.

① 植株 ② 叶 ③ 花蕾 ④ 花序 ⑤ 基生叶

川续断科 Dipsacaceae

蓝盆花属 *Scabiosa* Linn.

华北兰盆花 *Scabiosa tschiliensis* Gruün.

〔特征〕

多年生草本，根粗壮，木质。茎斜升，高20 ~ 50（80）cm。基生叶椭圆形，矩圆形，卵状披针形至窄卵形，先端略尖或钝，缘具缺刻状锐齿，或大头羽状裂，上面几光滑，下面稀疏或仅沿脉上被短柔毛，边缘具纤毛，叶柄长4 ~ 12cm；茎生叶羽状分裂，裂片2 ~ 3裂或羽裂，最上部叶羽裂呈条状披针形，长达3cm，顶端裂片长6 ~ 7cm，先端急尖。头状花序在茎顶成三出聚伞排列，直径3 ~ 5cm，总花梗长15 ~ 30cm，总苞片14 ~ 16片，条状披针形；边缘花较大而呈放射状；花萼5齿裂，刺毛状；花冠蓝紫色，筒状，先端5裂，裂片3大2小；雄蕊4；子房包于杯状小总苞内。果序椭圆形或近圆形，小总苞略呈四方柱状，每面有不甚显著中棱1条，被白毛，顶端有干膜质檐部，檐下在中棱与边棱间常有8个浅凹穴；瘦果包藏在小总苞内，其顶端具宿存的毛状萼针。花期6—8月，果期8—10月。

〔生境〕

生于沙质草原、典型草原及草甸草原群落中。

〔用途〕

药用，能清热泻火。

〔Features〕

Perennial herb, roots stout, woody. Stems obliquely ascending, 20 ~ 50 (80)cm tall. Basal leaves oval, oblong, ovate-lanceolate to narrowly ovate, slightly acute or obtuse at the apex, erose denticulate, or pinnately divided at the margin, glabrescent at the top, sparsely or short pubescent only along veins, ciliiform at the margin, petioles 4 ~ 12cm long; cauline leaves pinnatifid, lobes 2 ~ 3-lobed or pinnatifid, leaves pinnatifid and striped-lanceolate in the uppermost, up to 3cm long, lobes 6 ~ 7cm long at the apex, acuminate at the apex. Capitellate inflorescences ternately cymose arranged at the top of stems, 3 ~ 5cm in diameter, peduncles 15 ~ 30cm long, phyllary 14 ~ 16, striped-lanceolate; flowers larger and radiate at the margin; calyx 5 dentate fissure, setaceous; corollas blue purple, tubular, apex 5-lobed, 3 big lobes and 2 small; stamens 4; ovary wrapped in cyathiform involucel. Infructescence oval or suborbicular, involucel slightly four-square column shape, each face has inconspicuous 1 ridge, white pilose, scarious entablature at the apex, retuse depressions between middle ridge and marginal ridge below the entablature; achenes wrapped in the involucel, persistent hairy calyx needles at the apex. Flowering June to August, and fruiting August to October.

〔Habitat〕

Sandy grasslands, typical grasslands and meadow steppes.

〔Use〕

Medicinal use, clearing heat and discharging fire.

① 植株 ② 叶 ③ 花序 ④ 花蕾

桔梗科 Campanulaceae

沙参属 *Adenophora* Fisch.

长柱沙参 *Adenophora stenanthina* (Ledeb.) Kitag.

〔特征〕

多年生草本，高60～80cm，根粗大，长胡萝卜形。茎直立，上部分枝，有腺毛。基生叶有柄，卵形或近圆形；茎生叶互生，无柄，条状披针形或条形，边缘全缘或微齿，沿叶脉被柔毛。秋季开花，圆锥状花序或总状花序顶生，总花梗细长，花序倒挂状；花萼裂片五枚，无毛，全缘，条形及狭披针形之间，比花冠短；花冠钟状，5浅裂，蓝色或灰蓝色；花盘具短毛；雄蕊5，花丝下部变宽，有柔毛；子房下位，3室；花柱顶端三裂，花柱伸出。蒴果狭矩圆形，开裂；种子卵形，淡黄色。

〔生境〕

多生于山野草丛、山坡草地、草原；是山地草甸中最常见的伴生植物。

〔用途〕

药用，具有养阴清肺功效。

〔Features〕

Perennial herb, 60～80cm tall, roots thick, long carrot-shaped. Stems erect, branched at the top, with glandular hairs. Basal leaves petiolate, ovate or suborbicular; cauline leaves alternate, impedicellate, striped-lanceolate or striped, the margin entire or denticulate, pubescent along veins. Blooming in autumn, conical inflorescences or racemes terminal, peduncles elongated, inflorescence upside-down; calyx lobes 5, glabrous, the margin entire, striped and narrowly lanceolate, shorter than corolla; corolla campaniform, 5-lobed, blue or grey blue; flower disc short-pilose; stamens 5, filaments broadened at lower part, pubescent; ovary inferior, 3 locules; stylus trilobate at the apex, stylus extended. Capsules narrowly oblong, cracked; seeds ovate, light yellow.

〔Habitat〕

Wild grass, hillside meadows, grasslands; the most commonly seen associated plant on mountain meadows.

〔Use〕

Medicinal use, nourishing yin to clear away the lung-heat.

① 植株 ② 茎、叶 ③ 穗

菊 科 Compositae

蒿 属 *Artemisia* L.

冷 蒿 *Artemisia frigida* Willd.

〔特征〕

多年生半灌木状草本，丛生，具蒿臭味，灰白色，多数木质化短枝，根状茎横生。茎高20～40cm，稀匍匐分枝，密被灰白色绵毛。叶互生，无柄，二回羽状细裂，密被灰白色绵毛，裂片条形或披针状条形，长2～5mm，宽0.5～1mm，先端尖，全缘，茎上部叶不分裂。头状花序的半径0.2～0.7cm，多数，花序梗长4～5mm，于茎、枝上部排列成圆锥状；总苞片半球形，苞片1～4层，苞片椭圆形或宽披针形，密被灰白绵毛；花管状，两性，黄色，长0.2～0.3cm，花冠筒上部具3～5裂片；雄蕊5，花丝合生；子房下位，柱头两裂；花托密被白色长毛。瘦果圆柱形。

〔生境〕

生于草原、砾质山坡草地、荒地等。

〔用途〕

饲用；药用。

〔Features〕

Perennial suffruticose herb, fasciculate, artemisia smell, grey white, numerous lignified brachyblasts, rhizomes amphitropous. Stems 20～40cm tall, sparse decumbent branches, densely grey white lanate. Leaves alternate, impedicellate, bipinnate pinnate finely fissus, densely grey white lanate, lobes striped or lanceolate striped, 2～5mm long, 0.5～1mm wide, acute at the apex, the margin entire, leaves non-splitting at the top of stems. The radius of capitulums 0.2～0.7cm, numerous, inflorescence peduncles 4～5mm long, arranged into coniform at the top of stems and branches; phyllary hemispherical, bracts 1～4 layers, bracts oval or broadly lanceolate, densely grey white lanate; flowers tubular, bisexual, yellow, 0.2～0.3cm long, 3～5 lobes at the top of corolla tube; stamens 5, filaments concrescent; ovary inferior, stigmas bifid; receptacles are densely covered with white long-pilose. Achenes cylindrical.

〔Habitat〕

Grasslands, gravelly hillside meadows and wastelands, etc.

〔Use〕

Forage use; medicinal use.

植株

艾　蒿　*Artemisia argyi* Lévl. et Van.

〔特征〕

多年生草本，高45～120cm。茎直立，具条棱，密被白灰色绵毛，茎中部以上分枝。叶互生；茎下部叶三回深裂或浅裂，有时小裂片再次细裂；茎上部叶披针形，稀边缘有缺刻；叶全部具叶柄，基部楔形，上面疏散白色细粒状腺点，下面密被灰白色长毛。头状花序顶生，具多数花，成圆锥状花序，直径2.5mm，长5mm；总苞片密被长毛，苞片4～5层，覆瓦状，外层苞片卵形，内层苞片匙状矩圆形；花托半球形，无毛，花全管状，褐黄色；花序边缘花雌性，中央花两性；雄蕊5，花丝合生；子房下位，一室胚珠一枚，柱头两裂，瘦果长圆状，光滑，无冠毛。

〔生境〕

生于潮湿平原草地、荒野、山坡草地。

〔用途〕

药用；亦可作农药。

〔Features〕

Perennial herb, 45～120cm tall. Stems erect, carinal, densely white grey lanate, branched above the middle of stems. Leaves alternate; leaves at lower part of stems tripinnately parted or lobed, sometimes lobelets finely fissus; leaves at the top of stems lanceolate, sparsely erose at the margin; leaves have petioles, cuneate at the base, scattered white fine-grannular glandular dots at the top, densely grey white long-pilose beneath. Capitulums terminal, many-flowered, conical inflorescence, 2.5mm in diameter, 5mm long; phyllary densely long-pilose, bracts 4～5 layers, imbricate, outer bracts ovate, inner bracts spatulate oblong; receptacle hemispherical, glabrous, flowers tubular, brown yellow; female flowers at the margin of inflorescences, central flowers bisexual; stamens 5, filaments concrescent; ovary inferior, unilocular 1 ovule, stigmas bifid, achenes oblong-elliptic, glabrous, with no pappus.

〔Habitat〕

Wet plain grasslands, the wilds and hillside meadows.

〔Use〕

Medicinal use; it can also be used as pesticide.

① 植株
② 叶
③ 茎秆与头
④ 花序
⑤ 花序

蓝刺头属 *Echinops* L.

蓝刺头 *Echinops latifolius* Tausch

〔特征〕

多年生草本，茎直立，坚硬，下部密被褐色毛。叶大，宽矩圆状，边缘羽状深裂，上面绿色，密被蛛丝状毛；背面灰白色，密被蛛丝状毛，先端尖，基部半抱茎，裂片宽矩圆形或披针形，有时披针状三角形，顶端尖有针刺，边缘有不整齐缺刻或浅齿裂，裂片顶端有针刺，别的边缘有细刺。复合头状花序球形，直径6～8cm，顶生，多数单花头状花序组成，淡兰或深蓝色；总苞片外层为线状，中央苞片先端具细刺，内层苞片宽条形或披针形，边缘被柔毛。花管状，两性，淡蓝色，先端5裂，雄蕊5，紫蓝色花丝合生，子房下位，柱头两裂。瘦果长圆形或倒卵形，冠毛多数。

〔生境〕

生于干燥草原、山坡草地等。

〔用途〕

药用。

〔Features〕

Perennial herb, stems erect, hard, densely brown hairs beneath. Leaves large, broadly rounded-rectangular, pinnatiparted at the margin, green at the top, dense arachnoid hairs; greyish white on the back, dense arachnoid hairs, acute at the apex, semi-amplexicaul at the base, lobes broadly rounded-rectangular or lanceolate, occasionally lanceate-triangular, stylose at the apex, irregular erose or lobed at the margin, lobes stylose at the apex, spinelets at the margin. Composite capitulum spherical, 6 ~ 8cm in diameter, terminal, numerous simple flowers composed of capitulum, light blue or dark blue; outer phyllaries linear, central bracts speculate at the apex, inner bracts broadly striped or lanceolate, pubescent at the margin. Flowers tubular, bisexual, light blue, 5-lobed at the apex, stamens 5, violet filaments concrescent, ovary inferior, stigmas bifid. Achenes long elliptic or obovate, pappus numerous.

〔Habitat〕

Dry grasslands and hillside meadows.

〔Use〕

Medicinal use.

① 植株 ② 花与花蕾 ③ 茎秆 ④ 株丛 ⑤ 花序

蝟菊属 *Olgaea* Iljin.

蝟 菊 *Olgaea lomonosowii* (Trautv.) Iljin

〔特征〕

多年生草本，高15～30cm，根粗壮，木质，暗褐色。茎直立，具纵沟棱，密被灰白色绵毛，不分枝或由基部与下部分枝，枝细，毛较稀疏。叶近革质，基生叶矩圆状倒披针形，长10～15cm，宽3～4cm，先端钝尖，基部渐狭成柄，羽状浅裂或深裂，裂片三角形、卵形或卵状矩圆形，边缘具不等长小刺齿，上面浓绿色，有光泽，无毛，叶脉凹陷，下面密被灰白色粘毛，脉隆起；茎生叶矩圆形或矩圆状倒披针形，向上渐小，羽状分裂或具齿缺，有小刺尖，基部沿茎下延成窄翅；最上部叶条状披针形，全缘或具小刺齿。头状花序较大，单生于茎顶或枝端；总苞碗形或宽钟形，长2.5～4cm，直径3～5cm，总苞片多层，条状披针形，先端具硬长刺尖，暗紫色，具中脉1条，背部被蛛丝状毛与微毛，边缘有短刺状缘毛，外层者短，质硬而外弯，内层者较长，直立或开展。管状花两性，紫红色，长20～25mm，狭管部长6～9mm，檐部长14～16mm；花冠裂片5，长约4mm，顶端钩状内弯，花药尾部结合成鞘状，包围花丝。瘦果矩圆形，长5mm，稍扁，基部着生面稍歪斜；冠毛污黄色，不等长，长达22mm，基部结合。花果期8—9月。

〔生境〕

生于山谷、山坡、沙窝或河槽地。

〔用途〕

药用。

〔Features〕

Perennial herb, 15 ~ 30cm tall, roots stout, woody, dark brown. Stems erect, longitudinal furrows, densely grey white lanate, unbranched or branched at the base and lower part, branches slender, sparsely pilose. Leaves nearly leathery, basal leaves rounded-rectangular oblanceolate, 10 ~ 15cm long, 3 ~ 4cm wide, obtuse at the apex, tapering into petioles at the base, pinnately lobed or parted, lobes triangle ,ovate or oval-oblong, unequal-length deciduous teeth at the margin, strong green, lustrous, glabrous, veins concave at the top, densely grey white sticky hairs beneath, veins ridgy; cauline leaves oblong or rounded-rectangular oblanceolate, decrescent upward, pinnatifid or dissected, barbed, decurrent to narrow wings along stems at the base; leaves striped-lanceolate at the top, the margin entire or deciduous teeth. Capitulum slarger, solitary at the top of stems or the ends of the branches; involucres calathiform or short claws, 2.5 ~ 4cm long, 3 ~ 5cm in diameter, phyllary multi-layers, striped-lanceolate, hard and long barbed at the apex, dark purple, midrib 1, dorsal surface covered with arachnoid hairs and puberulous, short spiny tricholoma at the margin, outer ones short, hard and recurved, inner ones long, erect or squarrose. Tubular flowers bisexual, aubergine, 20 ~ 25mm long, narrow tube about 6 ~ 9mm long, entablature 14 ~ 16mm long; corolla lobes 5, about 4mm long, uncinate and recurved at the apex, combined into sheath-like at the tail of anthers, encircling filaments. Achenes oblong, about 5mm long, slightly flat, areole slightly oblique at the base; pappus dirty yellow, unequal length, up to 22mm long, connective at the base. Flowering and fruiting August to September.

〔Habitat〕

Valleys, hillsides, sand nests or river channels.

〔Use〕

Medicinal use.

① 植株 ② 分枝顶部

鳍蓟 Olgaea leucophylla (Turcz.) Iljin

〔特征〕

多年生草本，高15～70cm。茎直立，坚硬，具纵棱，密被白毛。叶互生，无柄，矩圆状披针形，长5～20cm，顶端具针刺，边缘羽状浅裂或具缺刻，裂片边缘具不等长的针刺，基部抱茎；叶上面疏生白毛，背面密被白毛，具白色。头状花序顶生，钟状；总苞片十多层，长约3.5cm，宽3～5cm，苞片披针形，深紫色，边缘具针刺，花全部管状，两性，粉红色，清香；花冠5裂，先端有钩刺；花托有毛。瘦果扁长圆柱状，先端短尖，基部收缩，黄褐色，无毛；冠毛深褐色，较密，不等长，基部合生。

〔生境〕

生于草地、沙质土壤、山坡草地等。

〔用途〕

药用。

〔Features〕

Perennial herb, 15～70cm tall. Stems erect, rigid, with longitudinal ridges, densely white pilose. Leaves alternate, impedicellate, rounded-rectangular lanceolate, 5～20cm long, stylose at the apex, pinnatilobate or erose at the margin, lobes with equilong stylose at the margin, amplexicaul at the base; the top of leaves with sparse white hairs, densely white pilose on the back, white. Capitulum terminal, campaniform; phyllaries a dozen lamellas, about 3.5cm long, 3～5cm wide, bracts lanceolate, dark violet, stylose at the margin, all flowers tubular, bisexual, pink, faint scent; corolla 5-lobed, barbed at the apex; receptacles pilose. Achenes prolate cylindrical, mucroniform at the apex, constricted at the base, yellowish brown, glabrous; pappus puce, dense, not equilong, concrescent at the base.

〔Habitat〕

Lawns, sandy soil and hillside meadows.

〔Use〕

Medicinal use.

① 植株 ② 花 ③ 花蕾 ④ 叶 ⑤ 茎秆

毛连菜属 Picris Linn.

毛连菜 *Picris hieracioides* Linn

〔特征〕

二年生草本，高16～120cm。根垂直直伸，粗壮。茎直立，上部伞房状或伞房圆状分枝，有纵沟纹，被稠密或稀疏的光亮分叉的钩状硬毛。基生叶花期枯萎脱落；下部茎叶长椭圆形或宽披针形，长8～34cm，宽0.5～6cm，先端渐尖或急尖或钝，边缘全缘或有尖锯齿或大而钝的锯齿，基部渐狭成长或短翼柄；中部和上部茎叶披针形或线形，较下部茎叶小，无柄，基部半抱茎；最上部茎小，全缘；全部茎叶两面特别是沿脉被亮色的钩状分叉的硬毛。头状花序较多数，在茎枝顶端排成伞房花序或伞房圆锥花序，花序梗细长。总苞片柱状钟形，长达1.2cm；总苞片3层，外层线形，短，长2～4mm，宽不足1mm，顶端急尖，内层长，线状披针形，长10～12mm，宽约2mm，边缘白色膜质，先端渐尖；全部总苞片外面被硬毛和短柔毛。舌状小花黄色，冠筒被白色短柔毛。瘦果纺锤形，长约3mm，棕褐色，有纵肋，肋上有横皱纹。冠毛白色，外层极短，糙毛状，内层长，羽毛状，长约6mm。花果期6—9月。

〔生境〕

生于山坡草地、林下、沟边、田间、撂荒地或沙滩地。

〔用途〕

药用。

〔Features〕

Biennial herb, 16 ~ 120cm tall. Roots vertical straight, stout. Stems erect, the upper part corymbose or branched, with longitudinal furrows, densely or sparsely shinning and forky uncinate strigose. Basal leaves withered and deciduous in the flowering phase; leaves of lower stems elliptic or broadly lanceolate, 8 ~ 34cm long, 0.5 ~ 6cm wide, acuminate or abruptly acute or obtuse at the apex, the margin entire or shorrp serrate or big and obtuse serrate, attenuate or brevialate at the base; leaves at middle and upper part of stems lanceolate or linear, smaller than lower leaves, impedicellate, semi-amplexicaul at the base; stems at the top small, the margin entire; all stem leaves clothed with uncinate bristles on both sides. Capitulum numerous, arranged into corymb or panicle at the top of stems, peduncles slender. Phyllaries campaniform, up to 1.2cm long; phyllaries 3 lamellas, outer ones linear, short, 2 ~ 4mm long, less than 1mm wide, acute at the apex, inner ones long, linear-lanceolate, 10 ~ 12mm long, about 2mm wide, white membranous at the margin, acuminate at the apex; all phyllaries strigose and short pubescent. Ligulate florets yellow, corolla tube white and short pubescent. Achenes spindle-shaped, about 3mm long, brown, with longitudinal ribs, rugose on ribs. Pappus white, outer ones very short, hispid, inner ones long, penniform, about 6mm long. Flowering and fruiting June to September.

〔Habitat〕

Hillside meadows, forests, groove edges, fields, abandoned lands or sand beaches.

〔Use〕

Medicinal use.

① 植株 ② 花蕾 ③ 叶丛 ④ 花

线叶菊属 *Filifolium* Kitam

线叶菊 *Filifolium sibiricum*(L.)Kitam

〔特征〕

多年生草本，茎基部残留厚密的纤维状叶鞘。基生叶莲座状，茎生叶互生，羽状全裂，末次裂片丝形。头状花序盘状，在茎枝顶端排列成伞房花序，边花雌性，1层，能育；盘花两性，通常不育；总苞半球形；总苞片3层，覆瓦状排列，无毛，卵形至宽卵形，边缘膜质，背部厚硬。花托稍凸起，蜂窝状。雌花花冠扁筒状，顶端稍收狭，2～4裂；两性花花冠筒状，顶端5裂，无狭管部。花柱2裂，顶端截形，花药基部钝，顶端有三角形附片。瘦果球状倒卵形，稍压扁，腹面有2条纹，无冠状冠毛。

〔生境〕

生于山坡、草地、山地及丘陵石质地。

〔用途〕

药用。

〔Features〕

Perennial herb, stems with residual thick sheath at the base. Basal leaves rosulate, cauline leaves alternate, pinnatisect, lobes filate. Capitulum discoid, arranged into corymb at the top of stems, ray flowers female, 1lamella, fertile; discoid flowers bisexual, usually sterile; involucre hemispherical; phyllaries 3 lamellas, imbricate, glabrous, ovoid to broadly ovoid, membranous at the margin, thick and hard on the back. Receptacles slightly convex, cellular. Female corollas flat tubular, slightly narrow at the apex, 2～3(4) lobed; bisexual corollas tubular, the apex 5-lobed, no narrow tube. Stylus 2-lobed, truncate at the apex, anther obtuse at the base, triangular accessory piece at the apex. Achenes globose-obovate, slightly flattened, 2 striped on the ventral side, no coronal pappus.

〔Habitat〕

Hillside, lawns, mountain lands, hills and stony lands.

〔Use〕

Medicinal use.

① 植株 ② 花 ③ 叶 ④ 花蕾

翠菊属 *Callistephus* Cass

翠　菊 *Callistephus chinensis* (L.) Nees

〔特征〕

一年或二年生草本，高（15）30～100cm。茎直立，单生，有纵棱，被白色糙毛，基部直径6～7mm，或纤细达1mm，分枝斜升或不分枝，下部茎叶花期脱落或生存；中部茎叶卵形、菱状卵形或匙形或近圆形，长2.5～6cm，宽2～4cm，顶端渐尖，基部截形、楔形或圆形，边缘有不规则的粗锯齿，两面被稀疏的短硬毛，叶柄长2～4cm，被白色短硬毛，有狭翼；上部的茎叶渐小，菱状披针形、长椭圆形或倒披针形，边缘有1～2个锯齿，或线形而全缘。头状花序单生于茎枝顶端，直径6～8cm，有长花序梗。总苞半球形，宽2～5cm；总苞片3层，近等长，外层长椭圆状披针形或匙形，叶质，长1～2.4cm，宽2～4mm，顶端钝，边缘有白色长睫毛，中层匙形，较短，质地较薄，染紫色，内层苞片长椭圆形，膜质，半透明，顶端钝。雌花1层，在园艺栽培中可为多层，红色、淡红色、蓝色、黄色或淡蓝紫色，舌状长2.5～3.5cm，宽2～7mm，有长2～3mm的短管部；两性花花冠黄色，檐部长4～7mm，管部长1～1.5mm。瘦果长椭圆状倒披针形，稍扁，长3～3.5mm，中部以上被柔毛。外层冠毛宿存，内层冠毛雪白色，不等长，长3～4.5mm，顶端渐尖，易脱落。花果期：5—10月。

〔生境〕

生于山坡撂荒地、山坡草丛、水边或疏林阴处。

〔用途〕

观赏。

〔Features〕

Annual or biennial herb, (15) 30 ~ 100cm tall. Stems erect, solitary, with longitudinal ridges, white hispid, the base 6 ~ 7mm in diameter, or slender as 1mm, branches obliquely ascending or eramose, lower stem leaves deciduous or persistent in the flowering phase; middle stem leaves ovoid, rhombic-ovoid or spathulate or suborbicular, 2.5 ~ 6cm long, 2 ~ 4cm wide, acuminate at the apex, truncate at the base, cuneate or rounded, irregularly serrate at the margin, sparsely hispidulous on both surfaces, petioles 2 ~ 4cm long, white hispidulous, with narrow wings; upper stem leaves decrescent, rhombic-lanceolate, oblong or oblanceolate, 1 ~ 2 serrated at the margin, or linear and the margin entire. Capitulum solitary at the top of stems, 6 ~ 8cm in diameter, with long peduncles. Involucre hemispherical, 2 ~ 5cm wide; phyllaries 3 lamellas, nearly equilong, outer lamellas oblong-lanceolate or spathulate, leaf texture, 1 ~ 2.4cm long, 2 ~ 4mm wide, obtuse at the apex, white ciliiform at the margin, middle lamellas spathulate, short, thin, purple, inner bracts oblong, membranous, semi-transparent, obtuse at the apex. Female flowers 1 lamella, multiple lamellas for gardening, red, light red, blue, yellow or bluish violet, ligulate 2.5 ~ 3.5cm long, 2 ~ 7mm wide, the tube 2 ~ 3mm long; bisexual corollas yellow, entablature 4 ~ 7mm long, the tube 1 ~ 1.5mm long. Achenes elliptoid oblanceolate, slightly flat, 3 ~ 3.5mm long, pubescent above the middle part. Outer pappus persistent, inner pappus snowy white, not equilong, 3 ~ 4.5mm long, acuminate at the apex, easily deciduous. Flowering and fruiting May to October.

〔Habitat〕

Hillside, wastelands, hillside grass land, waterfront or shade places of open forests.

〔Use〕

Ornamental value.

① 植株　② 花、叶　③ 花蕾

火绒草属　*Leontopodium* R.Brown

火绒草　*Leontopodium leontopodioides* Beauv.

〔特征〕

多年生草本，高15～25cm。茎丛生，直立或稍斜生，不分枝。全珠密生银灰色绵毛。基生叶丛生，茎生叶互生，无柄，披针形或条状被针形，全缘，顶端稍尖，两面密被毛。头状花序小，黄白色，3～5密集顶生，苞片2～3，条形披针形，不等长；边缘花雌性，花萼条形，花冠先端有小裂齿；中央花雄性或两性，管状，5裂；总苞片钟状，有多层苞片，密被毛，内层苞片披针形，先端膜质，外层苞片小，先端钝；花托无毛；雄蕊5，花药合生；子房下位；花柱不分裂或柱头微裂。瘦果矩圆形，长1mm；冠毛白色，基部合生。

〔生境〕

生于草原、山坡草地、溪边等地。

〔用途〕

药用。

〔Features〕

Perennial herb, 15～25cm tall. Stems fasciculate, erect or slightly oblique, unbranched. The whole plant is densely covered with silver grey lanate. Basal leaves fasciculate, cauline leaves alternate, impedicellate, lanceolate or striped aciculiform, the margin entire, slightly acute at the apex, both surfaces covered with dense hairs. Capitulums small, yellow white, 3～5 densely terminal, bracts 2～3, striped-lanceolate, unequal length; female flowers at the margin, calyx striped, corolla denticulate at the apex; central flowers male or bisexual, tubular, 5-lobed; phyllary campaniform, multi-layer bracts, covered with dense hairs, inner bracts lanceolate, membranous at the apex, outer bracts small, obtuse at the apex; receptacle glabrous; stamens 5, anthers concrescent; ovary inferior; stylus non-splitting or stigmas lobed. Achenes oblong, 1mm long; pappus white and concrescent at the base.

〔Habitat〕

Grasslands, hillside meadows and by the streams.

〔Use〕

Medicinal use.

① 植株 ② 花

狗娃花属 *Heteropappus* Less.

阿尔泰狗娃花 *Heteropappus altaicus*（Willd.）Novopokr.

〔特征〕

多年生草本，高10～50cm。须根多数簇生。茎斜升，有白色短毛，多分枝，有条棱。叶互生，无柄，条形或条状倒披针形，先端圆形或钝尖，基部楔形，边缘全缘。头状花序密集生于茎枝顶，边缘舌状花蓝紫色，雌性，中央管状花两性，黄色，花冠5深裂；总苞片半球形，有几层条状披针形小苞片；雄蕊5，花药复合，子房下位，一室，胚珠一枚。蒴果倒卵形，有短毛，有冠毛。

〔生境〕

生于丘陵、山坡草地、草原、路旁、河谷草地、砾质地。

〔用途〕

药用，具有温肺下气、化痰止咳功效。

〔Features〕

Perennial herb, 10～50cm tall. Fibrous roots fascicular. Stems obliquely ascending, white short-pilose, multi-branched, carinal. Leaves alternate, impedicellate, striped or striped-oblanceolate, rounded or obtuse at the apex, cuneate at the base, the margin entire. Capitulums densely aggregate at the top of stems, ligulate flowers blue purple at the margin, female, central tubular flowers bisexual, yellow, corolla 5-parted; phyllary hemispherical, several layers of striped-lanceolate bractlets; stamens 5, anthers composite, ovary inferior, unilocular, 1 ovule. Capsules obovate, covered with short-pilose and pappus.

〔Habitat〕

Hills, hillside meadows, grasslands, roadsides, valley grasslands and gravelly lands.

〔Use〕

Medicinal use, warming lung, preventing phlegm and relieving cough.

① 植株 ② 花 ③ 叶

橐吾属 *Ligularia* Cass.

全缘橐吾 *Ligularia mongolica* (Turcz.) DC.

〔特征〕

植株高30～80cm，全体呈灰绿色，无毛，茎直立，粗壮，直径3～10mm，具多数纵沟棱，常带紫红色，基部为褐色的枯叶纤维所包围。叶肉质，干后也较厚；基生叶矩圆状卵形、卵形或椭圆形，长6～20cm，宽2.5～8cm，先端钝圆，基部微心形，中部急狭而少下延至叶柄上，企业全缘或下部有波状浅齿，叶脉羽状，具长柄；基生叶2～3，椭圆形或矩圆形；中部叶有较短而下部抱茎的短柄；上部叶小，无柄而抱茎。头状花序在茎顶排列成总状，长可达25cm，多数，上部密集，下部渐疏离；花序梗在下部者长可达3cm，上部者长2～3mm；苞叶狭小，披针状钻形；总苞圆柱状，长10～13mm，宽3～5mm；总苞片5～6片，在外的矩圆状，条形，先端尖，在内的矩圆状倒卵形，先端钝，边缘宽膜质，背部有微毛；舌状花通常3～5，舌片短圆形，长15～20mm；管状花5～8，长约10mm。瘦果暗褐色，长约5mm。冠毛淡红褐色，长5～9mm。花果期7—8月。

〔生境〕

生于山地灌丛、石质坡地、沼泽草甸及林间。

〔用途〕

观赏，药用。

〔Features〕

The plant 30 ~ 80cm tall, all grey green, glabrous, stems erect, stout, 3 ~ 10mm in diameter, numerous longitudinal furrows, usually aubergine, base surrounded by brown fading leaves. Leaves succulent, thick when dry; basal leaves rounded-rectangular ovate, ovate or oval, 6 ~ 20cm long, 2.5 ~ 8cm wide, obtuse at the apex, slightly heart-shaped at the base, abruptly narrow and a few decurrent to petioles in the middle, the margin entire or undulate at lower part, veins pinnate, long petioles; basal leaves 2 ~ 3, oval or oblong; middle-part leaves have amplexicaul short petioles; leaves small at the top, impedicellate and amplexicaul. Capitulums arranged into racemous at the top of stems, up to 25cm long, numerous, dense at the top, lower part gradually alienated; peduncles as long as 3m beneath, 2 ~ 3mm long at the top; leaves narrow and small, lanceolate subulate; involucre cylindrical, 10 ~ 13mm long, 3 ~ 5mm wide; phyllaries 5 ~ 6, outer ones rounded-rectangular, striped, acute at the apex, inner ones rounded-rectangular obovate, obtuse at the apex, wide membranous margin, dorsal surface puberulous; ligulate flowers usually 3 ~ 5, ligulate short rounded, 15 ~ 20mm long; tubular flowers 5 ~ 8, about 10mm long. Achenes dark brown, about 5mm long. Pappus light russet, 5 ~ 9mm long. Flowering and fruiting July to August.

〔Habitat〕

Mountain lands, bushwoods, stony sloping fields, marsh meadows and forests.

〔Use〕

Ornamental value, medicinal use.

① 植株 ② 花 ③ 叶 ④ 花蕾

蓟 属 *Cirsium* Mill. Emend. Scop.

刺儿菜 *Cirsium setosum* (Willd.) MB.

〔特征〕

多年生直立草本，高50～100cm，有棱，被蛛丝状绵毛，上部多分枝，黄绿色。茎下部叶长椭圆形或矩圆形，基部楔形，顶端钝，边缘具缺刻状大裂或羽状浅裂和不等长的针刺，背面密被绵毛；茎上部叶较小，近全缘，披针形或卵形，边缘具浅裂和针刺。头状花序花多数，集生于茎的上部，枝顶部排列成伞房状花序；总苞片钟状，多层，内层苞片边缘膜质和紫色；初夏开紫红色花，花两性，管状5裂；雄蕊5，花丝合生；雌蕊1枚；子房下位；花柱柱头两裂，紫色，伸出花冠外，瘦果倒卵形，具四棱，冠毛星状。

〔生境〕

生于路旁、河边、平原、田地等潮湿地带。

〔用途〕

饲用；药用。

〔Features〕

Perennial erect herb, 50～100cm tall, carinal, arachnoid lanate, multi-branched at the top, yellow green. Leaves at lower part of stems oblong or rounded-rectangular, cuneate at the base, obtuse at the apex, erose dissected or pinnate lobed and unequal-length stylose at the margin, densely lanate on the back; leaves at the top of stems smaller, nearly the margin entire, lanceolate or ovate, lobed and stylose at the margin. Capitulums numerous flowers, aggregate at the top of stems, the top of branches arranged into corymbose; phyllaries campaniform, multilayers, inner bracts membranous margins and purple; aubergine flowers in early summer, flowers bisexual, tubular 5-lobed; stamens 5, filaments concrescent; pistil 1; ovary inferior; stylus stigma bifid, purple, stretching out of corolla, achenes obovate, tetragonous, pappus starlike .

〔Habitat〕

Roadsides, river banks, plains, fields and wetlands.

〔Use〕

Forage use; medicinal use.

① 植株 ② 花蕾 ③ 叶 ④ 花序 ⑤ 花

麻花头属 *Serratula* L.

麻花头 *Serratula centauroides* L.

〔特征〕

多年生草本，高50～70cm。茎直立，上部少分枝。基生叶丛生，具短柄，茎生叶互生，无柄，全长椭圆形，羽状深裂，长10～12cm，宽2～4cm，背面，疏生短糙毛，裂片条形或披针形，先端稍渐尖，边缘稀缺刻。头状花序顶生；总苞片钟状，苞片5层，直径1.5～2cm，外层苞片卵状披针形，长3mm，宽2mm，内层苞片狭披针形，长1.5cm，宽2～3mm，鳞片状，坚硬；花全管状，两性，花冠紫色，长2.5cm，基部窄，中间粗，先端5裂；花丝合生，但基部分离而箭形；花托被糙毛，瘦果光滑，圆柱形，长3mm；冠毛多数，黄褐色。

〔生境〕

生于荒地、山坡草地、平原等。

〔用途〕

药用，具有清热解毒、止血、止泻功效。

〔Features〕

Perennial herb, 50～70cm tall. Stems erect, subramous at the top. Basal leaves fasciculate, short petioles, cauline leaves alternate, impedicellate, oblong, pinnatiparted, 10～12cm long, 2～4cm wide, sparsely short hispid on the back, lobes striped or lanceolate, slightly acuminate at the apex, sparsely erose at the margin. Capitulums terminal; phyllaries campaniform, bracts 5 layers, 1.5～2cm in diameter, outer bracts ovate-lanceolate, 3mm long, 2mm wide, inner bracts narrowly lanceolate, 1.5cm long, 2～3mm wide, scaly, hard; flowers tubular, bisexual, corolla purple, 2.5cm long, narrow at the base, thick in the middle, apex 5-lobed; filaments concrescent, but separated and sagittal at the base; receptacle hispid, achenes glabrous, cylindrical, 3mm long; pappus numerous, yellowish-brown.

〔Habitat〕

Wastelands, hillside meadows and plains.

〔Use〕

Medicinal use, clearing heat and removing toxicity, arresting bleeding and antidiarrheic.

① 植株　2 花蕾　③ 叶　④ 花　⑤ 花蕾与花

苦苣菜属　*Sonchus* L.

苣荬菜　*Sonchus arvensis* L.

〔特征〕

多年生草本，高20～80cm。茎直立，具纵沟棱，无毛，下部常带紫红色，通常不分枝。叶灰绿色，基生叶与茎下部叶宽披针形、矩圆状披针形，长4～20cm，宽1～3cm，先端钝或锐尖，具小尖头，基部渐狭成柄状，柄基稍扩大，半抱茎，具稀疏的波状牙齿或羽状浅裂，裂片三角形，边缘有小刺尖齿，两面无毛；中部叶与基生叶相似，但无柄，基部多少呈耳状，抱茎；最上部叶小，披针形或条状披针形。头状花序多数或少数在茎顶排列成伞房状，有时单生，直径2～4cm，总苞钟状，长1.5～2cm，宽10～15mm，总苞片3层，先端钝，背部被短柔毛或微毛，外层着较短，长卵形，内层者较长，披针形；舌状花黄色，长约2cm。瘦果矩圆形，长约3mm，褐色，稍扁，两面各有3～5条纵肋，微粗糙；冠毛白色，长达12mm。花果期6—9月。

〔生境〕

生长于田间、村舍附近、路边。

〔用途〕

食用；饲用；药用，具有清湿热、消肿排脓、化瘀解毒功效。

〔Features〕

Perennial herb, 20 ~ 80cm tall. Stems erect, with longitudinal furrows, glabrous, lower part usually aubergine, and unbranched. Leaves grey green, basal leaves and leaves at lower part of stems broadly lanceolate, rounded-rectangular lanceolate, 4 ~ 20cm long, 1 ~ 3cm wide, obtuse or acute at the apex, sharply cuspidate, tapering into stipitiform at the base, the base of petioles slightly dilated, semi-amplexicaul, sparse undulate teeth or pinnatilobate, lobes triangle, barbed denticulate at the margin, glabrous on both surfaces; leaves in the middle part and basal leaves similar, but impedicellate, somewhat auriform at the base, amplexicaul; leaves at the top small, lanceolate or striped-lanceolate. Capitulums numerous or a few arranged into corymbose at the top of stems, occasionally solitary, 2 ~ 4cm in diameter, involucre campaniform, 1.5 ~ 2cm long, 10 ~ 15mm wide, phyllaries 3 layers; obtuse at the apex, dorsal surface short pubescent or puberulous, outer ones short, longovate, inner ones long, lanceolate; ligulate flowers yellow, about 2cm long. Achenes oblong, about 3mm long, brown, slightly flat, 3 ~ 5 longitudinal ribs on both surfaces, slightly coarse; pappus white, up to 12mm long. Flowering and fruiting June to September.

〔Habitat〕

Fields, cottage and roadsides.

〔Use〕

Edible use; forage use; medicinal use, eliminating dampness and heat, detumescent and expelling pus, removing blood stasis and detoxicating.

① 植株 ② 花序 ③ 叶 ④ 花 ⑤ 花蕾

蒲公英属 *Taraxacum* F. H. Wigg.

蒲公英 *Taraxacum mongolicum* Hand. - Mazz.

〔特征〕

多年生草本。根圆柱状，黑褐色，粗壮。叶倒卵状披针形、倒披针形或长圆状披针形，长4～20cm，宽1～5cm，先端钝或急尖，边缘有时具波装齿或羽状深裂，有时倒向羽状深裂或大头羽状深裂，顶端裂片较大，三角形或三角状戟形，全缘或具齿，每侧裂片3～5片，裂片三角形或三角状披针形，通常具齿，平展或倒向，裂片间常夹生小齿，基部渐狭成叶柄，叶柄及主脉常带红紫色，疏被蛛丝状白色柔毛或几无毛。花葶1至数个，与叶等长或稍长，高10～25cm，上部紫红色，密被蛛丝状白色柔毛；头状花序直径为30～40mm，总苞钟状，长12～14mm，淡绿色；总苞片2～3层，外层总苞片卵状披针形或披针形，长8～10mm，宽1～2mm，边缘宽膜质，基部淡绿色，上部紫红色，先端增厚或具小到中等的角状突起；内层总苞片线状披针形，长10～16mm，宽2～3mm，先端紫红色，具小角状突起；舌状花黄色，舌片长约8mm，宽约1.5mm，边缘花舌片背面具紫红色条纹，花药和柱头暗绿色。瘦果倒卵状披针形，暗褐色，长4～5mm，宽1～1.5mm，上部具小刺，下部具成行排列的小瘤，顶端逐渐收缩为长约1mm的圆锥至圆柱形喙长6～10mm，纤细；冠毛白色，长约6mm。花期4—9月，果期5—10月。

〔生境〕

广泛生于中、低海拔地区的山坡草地、路边、田野、河滩。

〔用途〕

药用，具有清热解毒、消肿散结功效。

〔Features〕

Perennial herb. Roots cylindrical, black brown, stout. Leaves obovate-lanceolate, oblanceolate or oblong-elliptic lanceolate, 4 ~ 20cm long, 1 ~ 5cm wide, obtuse or acuminate at the apex, occasionally undulate or pinnatiparted at the margin, sometimes runcinated-parted or pinnati-parted-lyrate, lobes larger at the apex, triangle or triangularly hastate, the margin entire or dentate, lobes 3 ~ 5 on each side, lobes triangle or triangular lanceolate, usually dentate, explanate or antitropous, denticulate among lobes, tapering into petioles at the base, petioles and primary veins usually reddish violet, sparsely arachnoid white pubescent or glabrescent. Scapes 1 to some, as long as leaves or slightly longer, 10 ~ 25cm tall, aubergine at the top, densely arachnoid white pubescent; Capitulums about 30 ~ 40mm in diameter, involucre campaniform, 12 ~ 14mm long, light green; phyllaries 2 ~ 3 lamellas, outer ones ovate-lanceolate or lanceolate, 8 ~ 10mm long, 1 ~ 2mm wide, wide membranous margin, light green at the base, aubergineat the top, apex incrassate or small and medium cornute protruding; inner ones linear lanceolate, 10 ~ 16mm long, 2 ~ 3mm wide, aubergine at the apex, small cornute protruding; ligulate flowers yellow, ligule about 8mm long, about 1.5mm wide, marginal ligule aubergine striped on the back, anthers and stigmas dark green. Achenes obovate-lanceolate, dark brown, about 4 ~ 5mm long, about 1 ~ 1.5mm wide, spinulate at the top, lower part with tactic tubercles, apex gradually shrunk into about 1mm-long conical to cylindrical with rostrum 6 ~ 10mm long, slender; pappus white, about 6mm long. Flowering April to September, and fruiting May to October.

〔Habitat〕

Widely found in mid and low altitude hillside meadows, roadsides, fields and beachlands .

〔Use〕

Medicinal use, clearing heat and removing toxicity, detumescent and removing stasis.

① 植株 ② 植株 ③ 叶 ④ 株丛

禾本科 Gramineae

芦苇属 *Phragmites* Adans.

芦　苇 *Phragmites australis*（Cav.）Trin. ex Steud.

〔特征〕

多年生草本。地下有粗壮匍匐的根状茎。秆长1～3m，直径2～10mm，节以下部位通常有白色粉霜。叶鞘圆筒状，无毛或被细毛；叶舌具毛；叶片扁平，长15～45cm，宽1～3.5cm，广披针形，光滑或边缘粗糙。圆锥花序长10～40cm，分枝稍伸展，下部枝腋被白毛；小穗含4～7小花，长12～16mm；颖具3脉，第一颖长3～7mm，第二颖长5～11mm；第一小花通常雄性，外稃长8～15mm，无毛，内稃长3～4mm，第二外稃长9～16mm，顶端长骤尖，具3脉，无毛，基盘具6～12mm长毛；内稃长3.5mm，背面有粗脊，颖果长圆柱形。

〔生境〕

生于池沼、河岸、道旁、湖泊边。

〔用途〕

药用，具有清胃火，除肺热功效；造纸；人造棉；编席、帘等用。

〔Features〕

Perennial herb.Strong rhizomes underground. Stalk is 1～3m long, 2～10mm in diameter, usually with white pruina on the lower part. The sheath cylindrical, glabrous and lanigerous; ligule pilose; leaf blade flat, 15～45cm long, 1～3.5cm wide, broadly lanceolate with smooth or rough edge. Panicle 10～40cm long, branches slightly stretched, lower axils white pilose; spikelets have 4～7 florets, 12～16mm long; glumes 3-veined, the first glume 3～7mm long, the second glume 5～11mm long; the first floret usually male; the lemma 8～15mm long, glabrous, palea 3～4mm long, the second lemma 9～16mm long, long and abruptly sharp at the apex with 3 veins, glabrous. Basal disc 6～12mm long pilose; palea 3.5mm long with ridge on the back. The caryopsis long cylindrical.

〔Habitat〕

Pond, river bank, roadsides or lake side.

〔Use〕

Medicinal use for clearing stomach fire and lung heat; papermaking; artificial cotton; matting and curtain, etc.

① 植株 ② 小穗

早熟禾属 *Poa* L.

草地早熟禾 *Poa pratensis* L.

〔特征〕

多年生，具发达的匍匐根状茎。秆疏丛生，直立，高50～90cm，具24节。叶鞘平滑或糙涩，长于其节间，并较其叶片为长；叶舌膜质，长1～2mm，蘖生者较短；叶片线形，扁平或内卷，长30cm左右，宽3～5mm，顶端渐尖，平滑或边缘与上面微粗糙，蘖生叶片较狭长。圆锥花序金字塔形或卵圆形，长10～20cm，宽3～5cm；分枝开展，每节3～5枚，微粗糙或下部平滑，二次分枝，小枝上着生3～6枚小穗，基部主枝长5～10cm，中部及下裸露；小穗柄较短；小穗卵圆形，绿色至草黄色，含3～4小花，长4～6mm；颖卵圆状披针形，顶端尖，平滑，有时脊上部微粗糙，第一颖长2.5～3mm，具1脉，第二颖长3～4mm，具3脉；外稃膜质，顶端稍钝，具少许膜质，脊与边脉在中部以下密生柔毛，间脉明显，基盘具稠密长绵毛；第一外稃长3～3.5mm；内稃较短于外稃，脊粗糙至具小纤毛；花药长1.5～2mm。颖果纺锤形，具3棱，长约2mm。花期5—6月，7—9月结实。

〔生境〕

生于湿润草甸、沙地、草坡。

〔用途〕

饲用；绿化，水土保持。

〔Features〕

Plants perennial with stout rhizomes, erect, sparsely clumped, 50 ~ 90cm long with 24 internodes. Sheath smooth or rough, longer than internodes and leaf-blades; ligute membranous, 1 ~ 2mm long, tiller shorter; leaf-blades linear, flat or involute, about 30cm long, 3 ~ 5mm wide, accuminate at the apex, smooth or slightly rough at the top, tillered leaf-blades long and narrow. Panicle pyramid-shaped or ovoid, 10 ~ 20cm long, 3 ~ 5cm wide; brachiate with 3 ~ 5 branches, slightly rough or smooth at lower part, secondary branches, 3 ~ 6 spikelets at each branchlet, main branches 5 ~ 10cm long at the base, middle and lower part exposed; stems of spikelets shorter; spikelets oval, green to grass green with 3 ~ 4 florets, 4 ~ 6mm long; glume oval-lanceolate, acute, smooth or slightly rough at the apex; the first glume 2.5 ~ 3mm long with 1 vein, second glume 3 ~ 4mm long with 3 veins; lemma membranous, slightly obtuse at the apex, densely pubescent below the middle part of ridge and marginal vein, conspicuous interstitial veins, basal disc dense and long lanose; first lemma 3 ~ 3.5mm long; palea shorter than lemma, ridge rough-ciliate; anthers 1.5 ~ 2mm long. Caryopsis spindle-shaped with trigonous, about 2mm long. Flowering May to June, fruiting July to September.

〔Habitat〕

Wet meadow, sandy land or grass slope.

〔Use〕

Forage use; greening, water and soil conservation.

① 植株 ② 小穗 ③ 景观

碱茅属 *Puccinellia* Parl.

星星草 *Puccinellia tenuiflora* (Griseb.) Scribn. et Merr.

〔特征〕

多年生，疏丛型。秆直立，高30～60cm，直径约1mm，具34节，节膝曲，顶节位于下部1/3处。叶鞘短于其节间，顶生者长5～10cm，平滑无毛；叶舌膜质，长约1mm，钝圆；叶片长2～6cm，宽1～3mm，对折或稍内卷，上面微粗糙。圆锥花序长10～20cm，疏松开展，主轴平滑；分枝2～3枚生于各节，下部裸露，细弱平展，微粗糙；小穗柄短而粗糙；小穗含2～3（4）小花，长约3mm，带紫色；小穗轴节间长约0.6mm；颖质地较薄，边缘具纤毛状细齿裂，第一颖长约0.6mm，具1脉，顶端尖，第二颖长约1.2mm，具3脉，顶端稍钝；外稃具不明显5脉，长1.5～1.8mm，宽约0.8mm，顶端钝，基部无毛；内稃等长于外稃，平滑无毛或脊上有数个小刺；花药线形，长1～1.2mm，花果期6—8月。

〔生境〕

生于草原盐化湿地、固定沙滩、沟旁渠岸草地。

〔用途〕

饲用。

〔Features〕

Plants perennial, sparse type. The stalk upright, 30～60cm tall, about 1mm in diameter with 34 internodes, geniculate. The epimerite located one third of lower part. Sheath shorter than its internodes, the terminal 5～10cm long, smooth and glabrous; ligule membranous, about 1mm long, obtuse rounded; leaf-blades 2～6cm long, 1～3mm wide, folded or slightly involute, rough at the top. Panicle 10～20cm long, spindle smooth; 2～3 branches born in each internode, lower part exposed, thin and flat, slightly rough; spikelet short and rough with 2～3 (4) florets, about 3mm long, purple; the internode length of apex about 0.6mm; glumes thinner, the edge ciliate serrate; the first glume about 0.6mm long with 1 vein, acute at the apex, the second glume about 1.2mm long with 3 veins, slightly obtuse at the apex; lemma 5 inconsipicuous veins, 1.5～1.8mm long, about 0.8mm wide, obtuse at the apex, glabrous at the base; palea as long as lemma, smooth glabrous or several small thorns on the ridge; anthers linear, 1～1.2mm lon. Flowering and fruiting June to August.

〔Habitat〕

Grassland salinized wetlands, fixed beaches, ditches or grassland.

〔Use〕

Forage use.

① 植株
② 成熟植株
③ 成熟期小穗
④ 开花期小穗
⑤ 茎、叶

雀麦属 *Bromus* L.

无芒雀麦 *Bromus inermis* leyss.

〔特征〕

多年生草本，根状茎横生。秆高45～80cm，直立，无毛或节以下部位有倒生刺。叶鞘无毛和只在鞘口分离；叶舌较硬，长1～2mm，叶披针形，长7～16cm，宽5～8mm，常无毛。圆锥状花序开展，长10～20cm，每节间分3～5枝，每枝具1～6小穗，小穗长12～25mm，含4～8小花；小穗柄具细纤毛；颖披针形，边缘膜质；外稃宽披针，无毛或基部稍糙，长8～11mm，具5～7脉，常无芒或脊上具1～2mm长芒；内稃比外稃短，脊上具毛；花药黄色，长3～4.5mm。颖果跟内稃等长。

〔生境〕

生于山坡草地、沟谷、路旁、河边。

〔用途〕

饲用。

〔Features〕

Perennial herb, rhizomes amphitropous. Stems 45～80cm tall, upright, glabrous or glochis at lower part of the internode. Sheath glabrous and only isolated at the sheath mouth; ligule hard, 1～2mm long, leaves lanceolate, 7～16cm long and 5～8mm wide, glabrous. The inflorescence is an open-branched panicle, 10～20cm long, 3～5 branches of each internode. Each branch has 1～6 spikelets. Each spikelet is 12～25mm long, containing 4～8 florets; spikelet has fine cilia; glume lanceolate, membranous margins; lemma broadly lanceolate, glabrous or slightly rough at the base, 8～11mm long, 5～7 veins, often unawned or 1～2mm long awned on the ridge; palea is shorter than lemma with hair on the ridge; anthers yellow, 3～4.5mm long. Caryopsis as long as palea.

〔Habitat〕

Hillside grasslands, valleys, roadsides and riverside.

〔Use〕

Forage use.

① 植株 ② 小穗 ③ 成熟期小穗

赖草属 *Leymus* Hochst.

羊　草 *Leymus chinense* (Trin.) Tzvel.

〔特征〕

多年生，具下伸或横走根茎；须根具沙套。秆散生，直立，高40～90cm，具4～5节。叶鞘光滑，基部残留叶鞘呈纤维状，枯黄色；叶舌截平，顶具裂齿，纸质，长0.5～1mm；叶片长7～18cm，宽3～6mm，扁平或内卷，上面及边缘粗糙，下面较平滑。穗状花序直立，长7～15cm，宽10～15mm；穗轴边缘具细小睫毛，节间长6～10mm，最基部的节长可达16mm；小穗长10～22mm，含5～10小花，通常2枚生于1节，或在上端及基部者常单生，粉绿色，成熟时变黄；小穗轴节间光滑，长1～1.5mm；颖锥状，长6～8mm，等于或短于第一小花，不覆盖第一外稃的基部，质地较硬，具不显著3脉，背面中下部光滑，上部粗糙，边缘微具纤毛；外稃披针形，具狭窄膜质的边缘，顶端渐尖或形成芒状小尖头，背部具不明显的5脉，基盘光滑，第一外稃长8～9mm；内稃与外稃等长，先端常微2裂，上半部脊上具微细纤毛或近于无毛；花药长3～4mm。花、果期6—8月。

〔生境〕

生于开阔平原、地山丘陵、河滩、盐渍低地。

〔用途〕

饲用。

〔Features〕

Perennial, rhizome stretching or horizontal; fibre roots with rhizosheath. Stalk scattered, erect, 40 ~ 90cm tall, 4 ~ 5 internodes. Sheath glabrous, residual sheath fibrous at the base, withered yellow; ligules truncate, carnassial at the top, papery, 0.5 ~ 1mm long; leaf-blades 7 ~ 18cm long, 3 ~ 6mm wide, flat or involute, coarse at the top and margin, and smooth beneath. Spike erect, 7 ~ 15cm long, 10 ~ 15mm wide; spike-stalk tiny ciliiform on the margin, internode 6 ~ 10mm long and 16mm long at the base; spikelet 10 ~ 22mm long, containing 5 ~ 10 florets, usually 2 born on 1 internode, or solitary at the apex and base, pink green, and turn yellow when mature; internode of spikelet glabrous, 1 ~ 1.5mm long; glume coniform, 6 ~ 8mm long, equal to or shorter than the first floret, not covering the base of the first lemma, hard with 3 veins, glabrous at the middle and lower part, coarse at the top, ciliiform at the margin; lemma lanceolate, the margin narrowly membranous, acuminate at the apex, dorsal surface has inconspicuous 5 veins, basal disc glabrous, the first lemma 8 ~ 9mm long; palea is the same long as lemma, bifid at the apex, fine ciliiform or glabrous at upper half of the ridge and anthers 3 ~ 4mm long. Flowering and fruiting June to August.

〔Habitat〕

Open plains, hills, floods, salt marsh and lowlands.

〔Use〕

Forage use.

① 植株
② 小穗

赖　草　*Aneurolepidium dasystachys* (Trin.) Nevski

〔特征〕

多年生草本，根状茎长，须根。秆直立，单生或疏丛，高45～90cm，具2～3节，上部密被柔毛，基部有许多老叶的残留物。叶长8～30cm，宽4～7mm，扁平或干时内卷。上面和边缘糙而疏生短毛。穗状花序直立，10～15cm，宽8～10cm，淡绿色；穗柄、小穗柄、外稃都被短毛；小穗长10～15mm，含4～7花，通常每节每小穗2～4枚；颖针刺状，具一脉，边缘有睫毛；每一颖比第二颖短，长8～12mm；外稃披针形，先端聚尖或具1～2mm长芒；基盘有约1mm长毛；内稃先端分裂状；花药长3.5～4mm。

〔生境〕

多生于芨芨草盐化草甸、马蔺盐化草甸、沙地、丘陵地、山坡、田间、路旁。

〔用途〕

饲用。

〔Features〕

Perennial herb, rhizome, fibrous root. Stalk erect, solitary or sparse scattered, 45～90cm tall, with 2～3 nodes, the top pubescent, a lot of residue at the base. Leaf is 8～30cm long, 4～7mm wide, flat or involute when dry. The top and margin coarse and sparsely pubescent. Spike erect, 10～15cm long, 8～10cm wide, light green; pedicel of spikelet and lemma short-pilose; spikelet 10～15mm long, containing 4～7 flowers, general each node has 2～4 spikelets; glume spinous, with one vein, ciliiform at the margin; the first glume shorter than the second one, about 8～12mm long; lemma lanceolate, sharp at the apex or 1～2mm-long awned; basal disc has 1mm long hairs; palea split at the apex; anthers about 3.5～4mm long.

〔Habitat〕

Widely spread on salinized meadows, saline meadows, sandy lands, hilly lands, hillside, field and roadsides.

〔Use〕

Forage use.

① 株丛 ② 开花期小穗 ③ 小穗

冰草属 *Agropyron* Gaertn.

冰 草 *Agropyron cristatum* (L.) Gaertn.

〔特征〕

多年生草本，须根稠密。秆疏丛生，上部被短毛，高30～60（75）cm，具2～3节，叶鞘紧抱茎，短于节间；叶片长5～20cm，宽2～5mm，较硬而粗糙，通常内卷。穗状花序较粗壮，长2.5～5.5cm，宽8～15mm；小穗紧密平行排列成两行，整齐，呈篦齿状，含4～7花，长10～13mm，宽约3mm；颖长方椭圆状；第一颖长2～3mm，第二颖长3～4mm，芒较长；外稃长6～7mm，边缘薄而窄，具短刺毛，芒长2～4mm；内稃于外稃基本等长，先端骤尖，背面主脉被短小刺毛；花药黄色，长约3mm。颖果长4mm。

〔生境〕

通常生于干燥草原，山坡、丘陵及沙地。

〔用途〕

饲用，药用。

〔Features〕

Perennial herb, dense fibrous root. Stalk sparsely fasciculate, short-pilose at the top, 30～60 (75)cm tall, 2～3 internodes, sheath stem-clasping, shorter than internodes; leaf-blades 5～20cm long, 2～5mm wide, hard and coarse, usually involute. Spike stout, 2.5～5.5cm long, 8～15mm wide; spikelet parallelly arranged in two lines, regular, pectinatory, containing 4～7 flowers, 10～13mm long, about 3mm wide; glume elliptic-oblong; the first glume 2～3mm long, the second glume 3～4mm long, long awn; lemma 6～7mm long, thin and narrow edge with bristles, awn 2～4mm long; palea as long as lemma, abruptly cuspidate at the apex, primary veins on the reverse side covered with short bristles; anthers yellow, about 3mm long. Caryopsis 4mm long.

〔Habitat〕

Usually grown on dry grasslands, hillsides, hills and sandy land.

〔Use〕

Forage use, medicinal use.

① 植株
② 小穗

洽草属 *Koeleria* Pers.

洽 草 *Koeleria cristata* (L.) Pers.

〔特征〕

多年生草本，密簇生，秆直立，高25～45cm，花序以下部位密被绒毛。叶鞘灰白或草黄，无毛或被短毛，在秆基碎裂呈纤维状；叶舌膜质，长0.5～2mm；叶片细条状，常内卷或扁平，灰绿色，长1.5～7cm，宽1～2mm，被短柔毛或上面无毛，边缘糙，基生枝上叶长5～15cm，宽约1mm。圆锥状花序紧缩呈穗状，长4～7cm，宽5～14mm，光滑，绿色或紫色，主轴和分枝有柔毛。小穗无毛，通常长4～5mm，有2～3（5）小花；颖倒卵状长圆形，边缘宽膜质，第一颖有1脉，长2.5～3.5mm，第二颖有3脉，长3～4.5mm；外稃披针形，具3脉，边缘膜质，无芒；内稃透明膜质，稍短于外稃，顶端两裂。花药长1.5～2mm。

〔生境〕

生于山坡草地、路旁、草地。

〔用途〕

饲用。

〔Features〕

Perennial herb, densely fasciated, stalk erect, 25～45cm tall, lower part of inflorescence densely tomentose. Sheath greyish white or straw yellow, glabrous or short-pilose, splintery and fibrous at the base of the stem; ligules membranous, 0.5～2mm long; leaf-blades thin striped, usually involute or flat, grey green, 1.5～7cm long, 1～2mm wide, covered with pubescence or glabrous at the top, coarse at the margin, leaves on base growing branches 5～15cm long, about 1mm wide. Conical inflorescence compressed and spicate, 4～7cm long, 5～14mm wide, glabrous, green or purple, main axis and branches have pubescence. Spikelets glabrous, usually 4～5mm long, containing 2～3(5) florets; glumes obovate-elliptic, wide membranous margins, the first glume with 1 vein, 2.5～3.5mm long, the second glume with 3 veins, 3～4.5mm long; lemmas lanceolate, with 3 veins, membranous margins, unawned; palea transparent membranous, slightly shorter than lemma, apex bifid. Anthers 1.5～2mm long.

〔Habitat〕

Hillsides, roadsides and meadows.

〔Use〕

Forage use.

① 植株 ② 花序 ③ 叶

拂子茅属 *Calamagrostis* Adans.

拂子茅 *Calamagrostis epigejos* (L.) Roth

〔特征〕

多年生草本，秆高45～100cm，具根状茎。叶鞘光滑或稍糙；叶舌膜质；叶片条形，扁平或内卷，长15～27cm，宽5～8mm，上面和边缘糙涩，背面光滑。圆锥花序密而狭，长10～20cm，常间断；小穗条状锥形，长5～7mm，灰绿色或微带紫色，含1小花；两颖等长或第二颖较短，草质，具一脉，或第二颖具三脉；外稃透明膜质，比颖短，先端齿裂，基盘之长柔毛于颖约等长；外稃背面中部或中部以上生2～3mm长芒；内稃比外稃短，先端具细缺刻；雄蕊3，花药长约1.5mm，黄色，子房卵形，柱头羽状两裂。

〔生境〕

生于低湿地、沙质土壤、干河谷内。

〔用途〕

饲用；编席；造纸原料。

〔Features〕

Perennial herb, stalk 45～100cm tall, with rhizomes. Sheath glabrous or slightly coarse; ligule membranous; leaf-blades striped, flat or involute, 15～27cm long, 5～8mm wide, coarse at the top and margin, glabrous at the reverse side. Panicle dense and narrow, 10～20cm long, disconnected; spikelet tapered striped, 5～7mm long, grey green or purply, 1 floret; two glumes equilong or the second glume shorter, herbaceous 1 veined, or the second glume 3 veined; lemma transparent membranous, shorter than glume, dentate fissure at the apex, the long pubescence of basal disc as long as glume; 2～3mm long awn grows on middle part or above of the reverse side of lemma; palea shorter than lemma, erose at the apex; 3 stamens, anthers about 1.5mm long, yellow, ovary ovoid, stigma pinnate, bifid.

〔Habitat〕

Low wetlands, sandy soil and dry valleys.

〔Use〕

Forage use; matting; raw material for paper making.

① 株丛 ② 种子、芒针、芒柱 ③ 种子、芒针、芒柱

针茅属 *Stipa* Linn.

大针茅 *Stipa grandis* P. Smirn.

〔特征〕

秆高50～100cm，具3～4节，基部宿存枯萎叶鞘。叶鞘粗糙或老时变平滑，下部者通常长于节间；基生叶舌长0.5～1mm，钝圆，缘具睫毛，秆生者长3～10mm，披针形；叶片纵卷似针状，上面具微毛，下面光滑，基生叶长可达50cm。圆锥花序基部包藏于叶鞘内，长20～50cm，分枝细弱，直立上举；小穗淡绿色或紫色；颖长3～4.5cm，尖披针形，先端丝状，第一颖具3～4脉，第二颖具5脉；外稃长1.5～1.6cm，具5脉，顶端关节处生1圈短毛，背部具贴生成纵行的短毛，基盘尖锐，具柔毛，长约4mm，芒两回膝曲扭转，微糙涩，第一芒柱长7～10cm，第二芒柱长2～2.5cm，芒针卷曲，长11～18cm；内稃与外稃等长，具2脉；花药长约7mm。花果期5—8月。

〔生境〕

多生于广阔、平坦的波状高原上。

〔用途〕

放牧；饲用。

〔Features〕

Stalk 50～100cm tall, 3～4 internodes, persistent withered sheath at the base. Sheath coarse or smooth when old, lower part usually longer than internodes; basal ligules 0.5～1mm long, obtuse, ciliiform at the margin, stalks 3～10mm long, lanceolate; leaf-blades vertically-rolled needle-like, puberulous on the top, glabrous underneath, basal leaves 50cm long. The base of panicle wrapped up in the sheath, 20～50cm long, branches thin, erect and scrolled; spikelets light green or purple; glumes 3～4.5cm long, lanceolate, filiform at the apex, the first glume 3～4 veins, the second glume 5 veins; lemma 1.5～1.6cm long, 5 veins, the joints a round of short-pilose at the apex, dorsal surface clothed with longitudinal short-pilose, basal disc acuminate with pubescence, about 4mm long, awn geniculate, twisted, slightly coarse, column length of the first awn 7～10cm and second awn 2～2.5cm, awn needle curly, 11～18cm long; palea as long as lemma, 2 veins; anther about 7mm long. Flowering and fruiting May to August.

〔Habitat〕

Most of them grow on vast and flat wavy highlands.

〔Use〕

Grazing; forage use.

① 植株 ② 花序 ③ 小穗 ④ 叶

芨芨草属 *Achnatherum* Beauv.

芨芨草 *Achnatherum splendens* (Trin.) Nevski

〔特征〕

多年生草本，根直径约3mm。秆多数，丛生，坚硬，高50～250cm，通常光滑，基部有许多老叶的残留物。叶鞘无毛，边缘膜质；叶片坚韧，纵向卷折，长30～60cm，上面有突出的叶脉，背面光滑。圆锥花序长40～60cm，夏季抽塔形灰绿色或带紫色圆锥花序；分枝细长，簇生，长约17cm；小穗长4.5～6.5mm；颖薄，先端尖，第一颖比第二颖短；外稃长4～5mm，具5脉，背面密被柔毛，先端两裂；基盘钝圆，无毛，长约0.5mm；芒不扭转而易断落，长5～10mm；内稃具两脉，无毛；花药长2.5～3mm，顶端有纤毛。

〔生境〕

通常生长在微碱性土壤中，分布于湖泊边草甸草原上。

〔用途〕

可作造纸，编织筐、篓、草席等；老株作冬季牧草。

〔Features〕

Perennial herb, root about 3mm in diameter. Most of stalk fasciculate, hard, 50～250cm tall, usually glabrous, a lot of residue of old leaves at the base. Sheath glabrous, membranous margins; leaf-blades tough and tensile, longitudinally convoluted, 30～60cm long, prominent veins on the top, glabrous on the back. Panicle 40～60cm long, turriform in summer, grey green or purply panicle; branches elongated, fascicular, about 17cm long; spikelet about 4.5～6.5mm long; glume thin, sharp at the apex, the first glume shorter than the second glume; lemma 4～5mm long, 5 veins, the back covered with pubescence, apex bifid; basal disc obtuse, glabrous, about 0.5mm long; awn not twisted but easily broken off, 5～10mm long; palea 2 veins, glabrous; anthers 2.5～3mm long, ciliiform at the apex.

〔Habitat〕

It usually grows in slightly alkaline soils and distributed in the meadow steppes by the river.

〔Use〕

It can be used for papermaking, woven crates, baskets, straw mats; old strains for winter forage.

① 植株 ② 花序 ③ 花序 ④ 叶

羽 茅 *Achnatherum sibiricum* (L.) Keng

〔特征〕

多年生草本，丛生。秆直立，光滑，高60～150cm，具3～4节；叶鞘疏松，光滑；叶片较硬，长20～60cm，宽3～7mm，通常卷折，背面光滑，上面和边缘粗糙。圆锥花序较密，长15～40（60）cm，分枝2～6簇生；小穗淡绿色，或变紫色，长8～10cm；颖膜质，无毛矩圆状披针形，等长或第二颖稍短；外稃长约7mm，先端微两裂，背面密被柔毛，基盘先端锐尖，长约1mm，密被毛；芒长约2.5cm，1回膝曲；内稃具2脉，有毛。颖果长圆柱形，长约4mm，正面有细槽。花果期6—9月。

〔生境〕

多生于干燥草原、山坡草地等。

〔用途〕

饲用。

〔Features〕

Perennial herb, fasciculate. Stalk erect, glabrous, 60～150cm tall, 3～4 internodes; sheath loose, glabrous; leaf-blades hard, 20～60cm long, 3～7mm wide, usually tortuous, glabrous on the back, coarse on the top and margin. Panicle dense, 15～40 (60)cm long, branches 2～6 fascicular; spikelet light green or turns to purple, 8～10cm long; glume membranous, glabrous rounded-rectangular lanceolate, equilong or the second glume slightly shorter; lemma about 7mm long, bifid at the apex, back covered with pubescence, basal disc acuminate at the apex, about 1mm long, covered with dense hairs; awn about 2.5cm long, geniculate; palea 2 veins, hairy. Caryopsis long cylindrical, about 4mm long with cannelure on obverse side. Flowering and fruiting June to September.

〔Habitat〕

Most of them grow on dry grasslands, hillside grasslands, etc.

〔Use〕

Forage use.

① 株丛顶拍 ② 株丛

隐子草属 *Cleistogenes* Keng

糙隐子草 *Cleistogenes squarrosa* (Trin.) Keng

〔特征〕

多年生草本，密丛生。秆高12～30cm，无毛，具多节，干时弯曲。叶鞘长于节间，无毛，花序以下层层抱茎；叶舌为一圈细纤毛；叶片通常内卷，粗糙，长3～6cm，基部宽1～2mm。圆锥花序细长，小穗少数，长4～7cm，宽5～10mm，每枝含2～5小穗；主轴、分枝、花序以下都粗糙；小穗长5～7mm，具2～3花，绿色或带紫色；颖通常具1脉，边缘宽膜质，无毛；外稃披针形，具3～5脉，边缘有毛，先端两裂，具短芒，基盘被短毛；内稃细长，于外稃等长或稍长，主脉伸出先端成约1mm长芒；花药2mm长。

〔生境〕

生于干草原、丘陵坡地、沙地、固定或半固定沙丘及山坡等处。

〔用途〕

饲用。

〔Features〕

Perennial herb, dense fasciculate. Stalk 12～30cm tall, glabrous, with multisections, curved when dry. Sheath longer than internodes, glabrous, amplexicaul layer upon layer below inflorescence; ligule has a round of fine ciliiform; leaf-blades usually involute, coarse, 3～6cm long, 1～2mm wide at the base. Panicle elongated, a few of spikelets, 4～7cm long, 5～10mm wide, each branch contains 2～5 spikelets; coarse below main axis, branches and inflorescence; spikelets 5～7mm long with 2～3 flowers, green or purply; glume usually 1 veined, wide membranous margin, glabrous; lemmas lanceolate, 3～5 veins, margin hairy, apex bifid, with short awn, basal disc short-pilose; palea elongated, as long as lemma or slightly longer, the central vein extending into a long spreading awn about 1mm long; anthers 2mm long.

〔Habitat〕

Dry steppes, hills and slopes, sandy lands, fixed or semi-fixed dunes and slopes.

〔Use〕

Forage use.

① 植株 ② 穗

狗尾草属 *Setaria* Beauv.

狗尾草 *Setaria viridis* (L.) Beauv.

〔特征〕

一年生草本。秆高（10）30～100cm，直立或基部膝曲，通常较细，有时基部粗4mm。叶鞘较疏松，无毛，稀有毛；叶舌长1～2mm，被细毛；叶片扁平，顶端骤尖，基部钝，长（3）5～30cm，宽2～15mm，通常无毛。圆锥花序密集成圆柱状，形似狗尾，长2～15(20)cm，夏季开花；小穗以下糙毛长4～12mm，粗糙，后变绿黄色或紫色，小穗椭圆形，先端钝，长2～2.5mm，通常三至数枚簇生于缩短的分枝上，具1～2花。第一颖卵形，长约小穗的1/3，具3脉；第二颖于小穗等长，和颖果约等长，具5（7）脉；第一外稃于小穗约等长，具5～7脉，内稃极细；第二稃鳞形背部突出而且抱内稃。颖果长圆柱形，先端钝，熟时于颖和外稃一起脱落。

〔生境〕

生于草原、路旁、田边。

〔用途〕

饲用；药用。

〔Features〕

Annual herb. Stalk (10) 30～100cm tall, erect or geniculate at the base, usually thiner, sometimes 4mm thick at the base. Sheath loose, glabrous, sparsely pilose; ligules 1～2mm long, lanulous; leaf-blades flat, abruptly cuspidate at the apex, obtuse at the base, (3 5～30cm long, 2～15mm wide, usually glabrous. Panicle crowded into cylindrical, dog-tail in form, 2～15 (20)cm long, bloom in summer; hispid 4～12mm long below spikelet, coarse, green-yellow or purple, spikelet oval, obtuse at the apex, 2～2.5mm long, usually 3 to more cluster on shortened branches, with 1～2 florets. The first glume ovate, about one third long of spikelet, with 3 veins; the second glume as long as spikelet, approximately equal in length to caryopsis, 5(7) veins; the first lemma approximately equal in length to spikelet, 5～7 veins, palea hairline; the second lemma squamous, dorsal surface extruding around the palea. Caryopsis long cylindrical, obtuse at the apex, and falls off with glume and lemma when mature.

〔Habitat〕

Grasslands, roadsides and fields.

〔Use〕

Forage use; medicinal use.

① 株丛 ② 花序 ③ 花序

菵草属

菵　草 *Beckmanning syzigachne* (Steud.)Fernald

〔特征〕

一年生草本，高40～70cm；秆基部微膝曲；叶片扁平，长6～13cm，宽2～7mm。圆锥花序窄，长15～25cm；小穗侧压扁，倒卵圆形至圆形，长3mm，含1小花，稀2小花；颖两侧压扁，背部较厚，边缘近膜质；外稃质薄，先端具芒尖，长约0.5mm；内稃等长于外稃或稍短。花期6—8月。

〔生境〕

生于湿地草甸、水边。

〔用途〕

饲用。

〔Features〕

Annual herb, 40～70cm tall; the base of stalk slightly geniculate; leaf-blades flat, 6～13cm long, 2～7mm wide. Panicle narrow, 15～25cm long; spikelets flattened, obovate to rounded, 3mm long, containing1 florets, sparsely 2 florets; glumes flattened on both sides, thicker on the back, nearly membranous at the margin; lemmas thin, aristate at the apex, about 0.5mm long; palea as long as lemmas or slightly shorter. Flowering June to August.

〔Habitat〕

Wet meadows and waterfront.

〔Use〕

Forage use.

① 植株 ② 花 ③ 叶 ④ 果实

百合科 LiLiaceae

知母属 *Anemarrhena* Bunge

知 母 *Anemarrhena asphodeloides* Bunge

〔特征〕

多年生草本，具匍匐根状茎，横生，长10～30cm，常半露于地面，外面密被黄褐色毛状叶鞘分裂物和须根。叶丛生，条形，稍硬，宽3～6mm，长20～70cm，无毛。花茎出自叶丛间，圆柱形，高50～100cm，外面被鳞片状小苞叶；总状花序顶生，通常2～3簇生，花白色，具淡紫色条纹，具有关节的短梗或无花梗；花被片6枚，两轮，长圆状线形，每片具三条紫色纵脉；雄蕊6，与内轮花被片对生；花丝极短，横生；子房卵形，三室。蒴果三角状卵圆形，有6条纵棱，每室有胚珠2粒；种子两端骤尖，具三棱，黑色。

〔生境〕

生于干燥丘陵草地、山坡、沙丘、草原及草甸草原。

〔用途〕

药用，具有清热、降火功效。

〔Features〕

Perennial herb, with decumbent rhizome, amphitropous, 10～30cm long, often semi-exposed on the ground, densely covered with yellowish-brown hairy sheath and fibrous root. Leaves fasciculate, striped, slightly hard, 3～6mm wide, 20～70cm long, glabrous. Rachis is from leafage, cylindrical, 50～100cm tall, clothed with scaly bracts on the outside; racemes terminal, usually 2～3 fascicular, floral white, with lavender striped, short peduncles or sessile; 6 tepals, 2 whorls, oblong-elliptic linear, each has 3 purple longitudinal veins; 6 stamens, opposite to inner tepals; filament very short, amphitropous; ovary ovoid, trilocular. Capsule triangular oval, 6 longitudinal ridges, each locule has 2 ovules; both ends of seeds cuspidate, trigonous, black.

〔Habitat〕

Dry hills, grasslands, hillsides, sand dunes, and meadow steppes.

〔Use〕

Medicinal use, clearing fire and downbear fire.

① 植株 ② 花蕾 ③ 叶、花蕾 ④ 小花与花蕾

萱草属 *Hemerocallis* L.

小黄花菜 *Hemerocallis minor* Mill.

〔特征〕

多年生草本。根状茎肉质，纺锤形。叶基生，条形，蓝绿色，无毛光滑，长30～60cm，宽5～10mm。花葶出自叶丛间，高40～80cm，无毛，花1～2（4）朵生于葶端，苞片被针形；花黄色，有香气，花常偏一侧生长，花梗长1.5～2cm；花被筒长1～2cm，卵球形，先端6裂，裂片长6～8cm；外轮裂片宽9～11mm，内轮裂片长椭圆形，边缘膜质，雄蕊6，比花被短；子房具三室，胚珠多数。蒴果椭圆形，长2.5～5cm，具三棱。

〔生境〕

生于草甸草原、山地草原。

〔用途〕

花蕾供食用；根入药，有毒，可作杀虫剂。

〔Features〕

Perennial herb. Rhizome succulent, fusiform. Leaves basal, striped, blue green, glabrous, 30～60cm long, 5～10mm wide. Peduncles out of leafage, 40～80cm tall, glabrous, 1～2(4) flowers born on the end of pedicels, bracts aciculiform; yellow flowers, aromatic, flowers usually inclined to one side, peduncles 1.5～2cm long; perianth tube 1～2cm long, ovoid, apex 6-lobed, lobes 6～8cm long; outer ones 9～11mm wide, inner ones long oval, membranous margins, stamens 6, shorter than perianth; ovary trilocular, many ovules. Capsule oval, 2.5～5cm long, trigonous.

〔Habitat〕

Meadow steppes, mountain lands and grasslands.

〔Use〕

Flower bud for edible use; roots can be used as medicine, poisonous, can be used as insecticide.

植株

葱　属　*Allium* L.

多根葱　*Allium polyrhizum* Turcz. ex Regel

〔特征〕

多年生草本，须根很多数而绳状。鳞茎在横生的根状茎上簇生，中央鳞茎枯萎，边缘鳞茎生长，圆柱形，直径约5mm，外皮黄色或深褐色丝状。茎高10～20cm，直立，稍具棱。叶片2～3枚，比茎短，很细长，半圆柱形，宽0.5～0.7mm。苞片白色，膜质；伞形花序半球形，花梗等长，比花被片长1.5倍；花被片6枚，紫红色，长圆状椭圆形，先端钝，长4～5mm；雄蕊6，伸出花被外；内层花丝基部有两个钝或尖形刻度；子房上位，三室；花柱细长，伸出花被外。蒴果比花被片稍短，种子扁平，黑色。

〔生境〕

生于平原草地、干燥草坡、沙质地、半荒漠地等。

〔用途〕

饲用；食用。

〔Features〕

Perennial herb, most of fibrous roots rope-form. Bulbs fascicular on the amphitropous rhizome, bulbs in the center withered, bulbs at the margin grow, cylindrical, about 5mm in diameter, cortex yellow or dark brown filiform. Stems 10～20cm tall, erect, slightly with ridges. 2～3 leaf-blades, shorter than stems, very elongated, semi-cylindrical, 0.5～0.7mm wide. Bracts white, membranous; umbels hemispherical, peduncles equilong, 1.5 times longer than tepals; 6 tepals, purple red, oblong-elliptic oval, obtuse at the apex, 4～5mm long; stamens 6, stretching out perianth; inner filaments have obtuse or apiciform marks at the base; ovary superior, trilocular; stylus elongated, stretching out of perianth. Capsule slightly shorter than tepals, seeds flat, black.

〔Habitat〕

Plains, grasslands, dry grass slopes, sandy lands and semi-deserts.

〔Use〕

Forage use; edible use.

① 植株 ② 小花 ③ 花序

野 韭 *Allium ramosun* L

〔特征〕

根状茎粗壮，横生。鳞茎近圆柱状，簇生，外皮暗黄色。叶三棱状条形，短于花葶。花葶圆柱状，高20～30cm，下部被叶鞘；总苞片单侧开裂，白色，膜质，宿存；伞形花序半球状或近球状，具多而较疏的花；小花梗近等长；花白色；花被片常具红色中脉；外轮花被片矩圆状卵形，先端具短尖头；内轮花被片矩圆状倒卵形；花丝等长，基部合生并与花被片贴生；子房倒圆锥状球形，具3圆棱，花柱不伸出花被外；子房外壁具疣状突起；蒴果小，倒卵状三角形，稀球形，顶端凹。种子黑色，具多棱。

〔生境〕

生于平原草地，沙质土壤及沙砾质坡地上。

〔用途〕

饲用。

〔Features〕

Rhizomes stout, amphitropous. Bulbs nearly cylindrical, fascicular, cortex dark yellow. Leaves trigonous striped, shorter than scape. Scape cylindrical, 20～30cm tall, bottom covered with sheath; phyllaries cracked, white, membranous, persistent; umbellate flowers hemispheric or nearly globose, with more and sparse flowers; florets and peduncles subequal; flowers white; tepals usually with red midribs; outer tepals rounded-rectangular ovoid, mucroniform at the apex; inner tepals rounded-rectangular obovate; filaments equilong, the base concrescent and adnate to tepals; ovary obconical spherical, 3 carinal, stylus not stretching out of perianth; ovary extine tuberculate; capsules small, obovate triangular, sparse spherical, concave at the top. Seeds black, carinal.

〔Habitat〕

Plain grasslands, sandy soil and gravelly sloping fields.

〔Use〕

Forage use.

① 植株 ② 花 ③ 叶

黄花葱 *Allium condensatum* Turcz.

〔特征〕

多年生草本。鳞茎单生，圆柱形，直径1～2cm，外皮纤维状。茎高30～80cm，无毛。圆柱状。叶4～7，生于茎基部，半圆柱形，具纵沟槽，实心，比茎短，无柄。总苞片白色膜质，比花序稍短；伞形花序球形，密生小花多数；花梗通常比花被长2～3倍，但是不超过3.5cm长；花被片6枚，黄色，细长卵形，长4～5mm，雄蕊6，黄色，生于花被基部，比花被片长；子房三角形，具三室。蒴果包裹在花被内，种子扁平，黑色。花果期7—8月。

〔生境〕

生于草原、潮湿草地、山坡草地、沟谷、沙砾质地。

〔用途〕

饲用；食用；药用。

〔Features〕

Perennial herb. Bulbs solitary, cylindrical, 1～2cm in diameter, cortex fibrous. Stems 30～80cm tall, glabrous, cylindrical. 4～7 leaves, grown on the basal part of stems, semi-cylindrical, with longitudinal furrows, solid, shorter than stems, impedicellate. Phyllary white membranous, slightly shorter than inflorescence; umbels spherical, densely florets; peduncles usually 2～3 times as long as perianth, but not exceeding 3.5cm long; 6 tepals, yellow, elongated, ovate, 4～5mm long, 6 stamens, yellow, born on the base of perianth, longer than tepals; ovary triangle, trilocular. Capsule wrapped up in perianths, seeds flat, black. Flowering and fruiting July to August.

〔Habitat〕

Grasslands, wet grasslands, hillside meadows, cheuches and gravel lands.

〔Use〕

Forage use; edible use; medicinal use.

① 植株 ② 花 ③ 花 4.花蕾 5.叶

鸢尾科 Iridaceae

鸢尾属 *Iris* L.

野鸢尾 *Iris dichotoma* Pall.

〔特征〕

多年生草本，高50～100cm。根状茎粗状而多节。茎黄绿色中空，多分枝；每枝下部有披针形苞片，苞片膜质，绿色。叶片剑形，排列成两行，蓝绿色，宽1.5～2.5cm，长4～30cm，顶端骤尖，基部套折状抱茎，边缘淡绿色全缘，无毛。花两性，由一鞘状苞内抽出，1～3朵顶生；花苞片长卵形，边缘膜质，长1.5～2cm；花被片6枚，两列，外轮花被片三枚，较大，长2.5～3cm，有爪，黄白色，中央面有一行鸡冠状黄褐色带紫纹突起；内轮三枚花被片较小，椭圆状，有钩状爪，蓝白色或蓝紫色，有紫色网纹；雄蕊3；花柱分枝三，柱头深裂，呈花瓣状并有美丽的颜色，子房下位，三室，胚珠多数。蒴果圆柱形，长3 ~ 5cm；种子多数，近球形，暗褐色，两端翅状。花果期7—9月。

〔生境〕

生于向阳坡和草甸草原。

〔用途〕

药用，具有活血祛瘀功效。

〔Features〕

Perennial herb, 50~100cm tall. Rhizomes thick and multi-sectional. Stems yellow green and hollow, multi-branched; the lower part of each branch has lanceolate bracts, bracts membranous, green. Leaf-blades sword-like, arranged in two rows, blue green, 1.5 ~ 2.5cm wide, 4 ~ 30cm long, cuspidate at the apex, amplexicaul at the base, light green at the margin, glabrous. Flowers bisexual, extracting from sheath buds, 1 ~ 3-flowered terminal; bracts long ovate, membranous margins, 1.5 ~ 2cm long; 6 tepals, 2 lines, 3 larger outer tepals, 2.5 ~ 3cm long, unguiculate, yellowish white, the median plane cristate yellowish-brown protruding with purple striae; 3 smaller inner tepals, elliptical, uncinate, blue white or blue purple, with purple striae; 3 stamens; stylus 3 branched, stigma parted, petaloid with beautiful colors, ovary inferior, trilocular, multiple ovules. Capsule cylindrical, 3 ~ 5cm long; multiple seeds, subsphaeroidal, dark brown, aliform at both ends. Flowering and fruiting July to September.

〔Habitat〕

Sunny lands and meadow steppes.

〔Use〕

Medicinal use, activating blood circulation and removing stasis.

① 植株
② 株丛
③ 果实
④ 花

细叶鸢尾 *Iris tenuifolia* Pall.

〔特征〕

多年生草本，高15～40cm，根状茎。茎基部有许多褐色丝状叶鞘的残留物。叶多数，基部丛生，丝状条形，长20～50cm，宽1～2.5mm，淡绿色，具5～7条纵脉，无毛。花葶由叶簇中抽出，细长，较短，有节；苞片3～4，条状披针形，淡黄绿色或有时紫色，边缘膜质，顶端骤尖，内层裂片长约8cm；花1～2朵，淡蓝紫色，长4～7cm；花被管长3～5cm，先端裂片细长；外轮花被片较大，先端淡黄白色或淡蓝色，内轮花被片蓝色；雄蕊3；子房下位，三室；雌蕊花柱细长，花瓣状，蓝色，外卷，宿存。蒴果球形或倒卵形，长约2cm，宽约1.5cm，具3～6棱；种子扁卵形，黄褐色。花期5月，果期6—7月。

〔生境〕

生于干燥草原、沙质地、荒野。

〔用途〕

药用，具有安胎养血功效。

〔Features〕

Perennial herb, 15～40cm tall, rhizome. Basal part of stem has many residues of brown filiform sheath. Multiple leaves, fasciculate at the base, filiform striped, 20～50cm long, 1～2.5mm wide, light green, 5～7 longitudinal veins, glabrous. Peduncles pulling out from leaves, elongated, short, jointed; 3～4 bracts, striped lanceolate, yellowish green or purple, membranous margins, cuspidate at the apex, inner lobes about 8cm long; 1～2 flowered, light blue purple, 4～7cm long; perianth 3～5cm long, lobes elongated at the apex; outer tepals larger, yellowish white or light blue at the apex, inner tepals blue; 3 stamens; ovary inferior, trilocular; pistil stylus elongated, petaloid, blue, revolute, persistent. Capsule spherical or obovate, about 2cm long, about 1.5cm wide, 3～6 ridges; seeds flat ovate, yellowish-brown. Flowering in May, and fruiting June to July.

〔Habitat〕

Dry grasslands, sandy lands and the wilds.

〔Use〕

Medicinal use, miscarriage prevention and nourishing blood.

① 花与叶 ② 花 ③ 近景 ④ 远景

马 蔺 *Iris pallisii* Fisch.var. *chinensis* Fisch.

〔特征〕

多年生草本。根状茎短而粗壮。基部常具残叶裂成的纤维状毛，基生叶丛生。叶条形，宽3～8mm，绿色或深绿色，基部稍红褐色并较坚韧。花葶高15～40cm，顶端生于1～3朵淡蓝紫色花；穗状苞片三个，叶式条形，长8～10cm，花被片6枚，基部稍合生，外轮花被片较大，匙形，中部有黄色条纹，往外弯曲下垂，内轮花被片倒披针形，直立；雄蕊3，花丝长而外卷；子房下位，三室，具三棱；花柱分枝三枚，花瓣状，蓝色，边缘有齿牙。蒴果长椭圆形，具3～6条纵棱；种子多数，红褐色，有棱。

〔生境〕

生于山野草地、山坡草原、河边、路旁、草甸草原。

〔用途〕

药用，具有止血、利尿功效；造纸原料。

〔Features〕

Perennial herb. Rhizomes short, stout. Fibrous hairs of leases at the base, basal leaves fasciculate. Leaves striped, 3～8mm wide, green or dark green, slightly russet and tough and tensile at the base. Peduncle 15～40cm tall, 1～3 blue purple flowers at the apex; 3 spicate bracts, leaf-form striped, 8～10cm long, tepals 6, slightly concrescent at the base, outer tepals larger, spathulate, yellow striped in the middle part, curved pendulous outward, inner tepals oblanceolate, erect; stamens 3, filaments long and revolute; ovary inferior, trilocular, trigonous; 3 stylus branches, petaloid, blue, denticulate at the margin. Capsule long oval, 3～6 longitudinal ridges; multiple seeds, russet, carinal.

〔Habitat〕

Wild grasslands, hillsides, river banks, roadsides and meadow steppes.

〔Use〕

Medicinal use, arresting bleeding, diuretic; papermaking raw material.

参考文献

陈山. 1994.中国草地饲用植物资源[M]. 沈阳:辽宁民族出版社.

Chen Shan. 1994. Forage Plant Resources of China's Grasslands[M]. Shenyang: Liaoning Nationalities Publishing House.

崔乃然. 1980.植物分类学[M]. 北京:农业出版社.

Cui Nairan. 1980. Plant Taxonomy[M]. Beijing: Agricultural Publishing House.

富象乾. 1990. 内蒙古饲用植物[M].呼和浩特：内蒙古人民出版社.

Fu Xiangqian. 1990. Forage Plants of Inner Mongolia[M].Hohhot: Inner Mongolia People's Publishing House.

耿以礼. 1959.中国主要植物图说-禾本科[M]. 北京:科学出版社.

Geng Yi-li. 1959. Flora illustralis Plantarum Primarum Sinicarum-Gramineae[M]. Beijing: Science Press.

谷安琳. 2009.中国北方植物彩色图谱[M].北京：中国农业科学技术出版社.

Cu Anlin. 2009. Atlas of Rangeland Plants in Northern China[M].Beijing: China Agricultural Science and Technology Press.

吉木色. 2008.内蒙古草原常见植物图鉴[M].呼和浩特：内蒙古人民出版社.

Ji Muse.2008. Atlas of Common Plants on Inner Mongolia Grassland[M]. Hohhot:Inner Mongolia People's Publishing House.

吉木色. 2010.正蓝旗草原常见植物图鉴[M].呼和浩特：内蒙古人民出版社.

Ji Muse.2010. Atlas of Common Plants on Zhenglan Banner Grassland[M]. Hohhot:Inner Mongolia People's Publishing House.

吉日木图, 萨日娜. 2004.生物学名词术语[M].呼和浩特：内蒙古教育出版社.

Jiri Mutu, Sanrina. 2004. Terminology of Biology[M]. Hohhot: Inner Mongolia Education Press.

蒋尤泉. 2007.中国作物及其野生近缘植物-饲用及绿肥作物篇[M].北京：中国农业出版社.

Jiang Youquan. 2007. Crop and Its Relative Wild Plants and Green Manure Crop in China[M]. Beijing: China Agricultural Press.

刘磊. 2015.内蒙古植物资源分类彩色图谱（一）[M].呼和浩特：内蒙古大学出版社.

Liu Lei.2015.Atlas of Plant Resources of Inner Mongolia (I) [M]. Hohhot:Inner Mongolia University Press.

马毓泉. 1998.内蒙古植物志[M]. 内蒙古人民出版社.

Ma Yuquan. 1998. Flora of Inner Mongolia[M]. Hohhot: Inner Mongolia People's Publishing House.

内蒙古师范学院，教育出版社自然科学编辑室. 1976.种子植物图鉴[M].呼和浩特：内蒙古教育出版社.

Inner Mongolia Normal University, Natural Science Editor Room. 1976. Illustrated Handbook of Seed Plants[M]. Hohhot: Inner Mongolia Education Press.

内蒙古植物志编辑委员会. 1989-1994.内蒙古植物志: 1-5卷 [M]. 第2版.呼和浩特：内蒙古人民出版社.

Editorial Committee of Folra of Inner Mongolia. 1989-1994. Flora of Inner Mongolia: Vol.1-5[M]. 2nd.Hohhot: Inner Mongolia Poprlar Press.

内蒙古植物志编写组. 1978-1985.内蒙古植物志: 1-8卷 [M].呼和浩特：内蒙古人民出版社.

Writing Group of Flora of Inner Mongolia. 1978-1985. Flora of Inner Mongolia (1st edition):Vol. 1-8[M]. Hohhot: Inner Mongolia People's Publishing House.

赵建成, 吴跃峰. 2002.生物资源学[M].北京：科学出版社.

Zhao Jiancheng, Wu Yuefeng. 2002. Biological Resources Science[M].Beijing: Science Press.

中国科学院北京植物研究所. 1976–1983.中国高等植物图鉴:1-5卷 [M]. 北京:科学出版社.

Institute of Botany. 1976–1983. Iconographia Cormophytorum Sinicorum:Vol.1-5 [M]. Beijing: Science Press.

中国科学院内蒙古宁夏综合考察队. 1985.内蒙古植被[M]. 北京:科学出版社.

Team of the Comprehensive Scientific Expedition to Inner Mongolia-Ningxia, Chinese Academy of Sciences. 1985. Vegetation of Inner Mongolia[M]. Beijing: Sciences Press.

周云龙. 1999 植物生物学[M].北京: 高等教育出版社.

Zhou Yunlong. 1999. Plant Biology[M]. Beijing: Higher Education Press.

朱家柟. 2001.拉汉英种子植物名称[M].北京：科学出版社.

Zhu Jianan. 2001. Dictionary of Seed-Plants Names-Latin-Chinese-English[M]. Beijing:Science Press.

拉丁名、中英文名索引

T

V